$41.75

Introduction
to Lasers and
Their Applications

Introduction to Lasers and Their Applications

DONALD C. O'SHEA
School of Physics, Georgia Institute of Technology

W. RUSSELL CALLEN and **WILLIAM T. RHODES**
School of Electrical Engineering, Georgia Institute of Technology

ADDISON-WESLEY PUBLISHING COMPANY

Reading, Massachusetts
Menlo Park, California · London · Amsterdam · Don Mills, Ontario · Sydney

This book is in the
ADDISON-WESLEY SERIES IN PHYSICS

Second printing, December 1978

ISBN 0-201-05509-0
MNO-MA-8987

To Hetch, Mary Anne, and Judy

Preface

The laser is now found not only in the research laboratory but in the automobile factory, on the construction site, and even in the supermarket. A great need exists for those outside the ranks of research scientists and engineers to have a broader familiarity with this recent addition to today's technology. Most important, we believe, are the students of the technical professions preparing for tomorrow's world and its challenges. The course on which this textbook is based is but one of a great number of modern device-oriented courses that are now being taught at today's colleges, universities, and institutes.

There presently exists a gap between the brief reviews of lasers provided in modern physical optics texts and the thorough, graduate-level texts on lasers and quantum mechanics. This book is designed to fill this gap and to serve as the basis for a short, self-contained course that gives the undergraduate a feel for modern laser technology. For those students who may not want to invest a substantial amount of their elective time in extensive course work in this area, it represents a reasonable alternative to a more lengthy treatment. The text is based upon a series of notes used in a one-quarter, three-credit-hour course at the Georgia Institute of Technology, open to any student who has passed the calculus-based physics sequence (text: Halliday and Resnick, *Fundamentals of Physics*, Wiley, 1974). The mathematical level of the text is kept relatively simple (a few integrals and some simple differential equations) to ensure that it will be useful to a large number of students, not just the mathematically more sophisticated physics and engineering students. Stated concisely, the objectives of the text are the following: (1) to give the student the concepts and vocabulary needed to understand laser advances and applications discussed in the technical magazines of the laser field (see references at the end of Chapter 1), and (2) to enable the student to understand a laser specification sheet. If these objectives are met, the student should be able to understand most laser applications in his or her field of specialization.

We expect the text to be of use to some who are not students. Laser technicians using the preliminary version of the text have found it useful. We have not avoided mathematical descriptions where they were needed, but would note that most of the math leads to a figure or an expression that gives an understandable result and serves as a point of continuation for the balance of the discussion. Not wishing to overwhelm the student, we have in places backed off from an exhaustive explanation. A teacher may feel it worthwhile to add additional material in these areas.

The text is organized into two general parts. The early portion (Chapters 1–4) emphasizes the physical theory needed to understand lasers; the latter portion (Chapters 5–9) emphasizes the devices themselves, their construction, and their applications. After a brief introductory chapter, the properties of laser light are discussed in some detail. We have chosen to begin with a discussion of these properties rather than with the conventional exposition on atomic energy levels, Einstein relations, and the concept of stimulated emission for two reasons. (1) Our approach allows a number of laser demonstrations at the beginning of the course, and (2) it provides a review and expansion of necessary physical optics principles that are referred to throughout the text. Particularly important are the discussions of interferometers and coherence. We realize that Chapter 2 is rather lengthy, but for those students who have taken a physical optics course as a prerequisite, it can be quickly reviewed.

The next three chapters constitute a spiral approach to understanding the laser. Chapter 3 provides a first look at the basic requirements of an idealized light amplifier and a laser. The idealization is dropped in Chapter 4, where the finite laser lineshape and gain saturation are discussed. This leads to a description of the output of a laser of unspecialized design. In Chapter 5, we examine modifications of the laser output obtained by introducing various devices into the laser cavity. This chapter is a transition chapter in that it is a mixture of physical theory and device description.

Chapter 6, a detailed discussion of various types of lasers, is almost completely device-oriented. We have chosen to describe those lasers that are most widely used or that illustrate an interesting point of laser theory. To be all-inclusive, we feel, would require too much introductory material for specialized laser systems. For example, the tunable spin-flip Raman laser is not discussed.

Chapters 7, 8, and 9 cover three areas of laser applications: holography, communications, and power. Just as with the types of lasers, we could not present the mass of material necessary to give adequate descriptions of all the laser applications known today. These three areas were chosen because they are individually important and because they are representative of the broad range of possible applications. Other areas were omitted for a number of reasons. Applications of lasers to alignment and precision measurement are for the most part easily grasped and need little exposition. The area of medical applications resembles that of power applications (focusing of a laser beam to cut or heat a small area), but would require considerable introduction of medical terminology and techniques. In the area of optical identification and scanning, the laser is used as a glorified light bulb of high brightness and directional output. The applications of lasers to spectroscopy, while of interest to many scientists, are so varied and require so much introductory preparation that only a very lengthy chapter would do justice to this application. Still, an instructor who is working in one of these laser-related fields could supplement or substitute for the three we have chosen those of his or her own interest. In addition, a strong student interest in a particular field should be taken into account. If this course is taught on a semester basis, the instructor should have no trouble filling the time available; additional applications can be added very easily.

In all demonstrations and experiments, the safety of the student and the instructor should be a major consideration. Rather than provide a separate chapter on laser safety, we have included cautions and discussions on laser safety within the body of the text. This makes for a series of continuing reminders instead of a single section that could easily be ignored. At this writing, the regulations for lasers in some states and at the federal level are still being debated. The instructor should keep informed of the restrictions on the use of lasers and pass this information on to his or her students.

We wish to acknowledge the help, patience, and criticisms of our students who have taken the laser course over the past five years. We also thank Earl McDaniel and Charles Braden of the faculty of the School of Physics, Georgia Institute of Technology, for critical readings of an early version of the notes. We are particularly indebted to Hugo Weichel, Charles Hathaway, David Burch, Clifford Fairchild, Murray Sargent, Stephen Jacobs, and Shaoul Ezekial for their excellent suggestions and criticisms of the manuscript. We appreciate the cooperation and support of James Stevenson, Director of the School of Physics, and Demetrius Paris, Director of the School of Electrical Engineering, Georgia Institute of Technology. Finally, having produced this book, we now understand why other authors always thank their spouses with such fervor. It is they who must put up with the greatest amount of hassle. Thank you, dear wives . . . you were great!

Atlanta, Georgia   D.C.O.
January 1977   W.R.C.
                                                                                                     W.T.R.

Contents

Lasers:
A Short Introduction

The laser is now approaching the end of its second decade. During this period, it has developed from what was once described as "an invention in search of an application"* into one of the most important technological developments of this half-century. It has brought about a rebirth of the science and technology of optics and has led to the development of whole new industries. Viewed from almost any perspective, the laser is a remarkable device. Consider:

- A laser time standard accurate to a small fraction of a second per year.
- A laser beam so directional it can easily be seen from the moon—or reflected back to earth and detected here.
- A glass bead supported in air only by a shaft of green light from a laser.
- Laser-based measuring systems so accurate that they determine the altitude of earth-orbiting satellites to within several meters; the surface deformations of vibrating objects to 0.05 nanometers (billionths of a meter).
- Focused laser beams so intense that they initiate nuclear reactions.
- Huge industrial lasers with many thousands of watts of output in beams the diameter of one's finger.

The variety of lasers and the wealth of laser applications developed since 1960 is enormous.

Short History of the Laser

Until 1917, no one conceived that there was a basic process that would allow light to be amplified as it is in a laser. In that year, Albert Einstein showed that the process of stimulated emission must exist, and from that time the invention

---

* A description applied by Dr. A. L. Schawlow, coinventor of the laser principle.

of the laser was possible. During the 1920's, 30's, and 40's, physicists were pre-occupied with the new discoveries of quantum mechanics, particle physics, and nuclear physics. For the most part, the possibility of laser action lay dormant, although the needs of science and technology for such a device grew. Besides the science-fiction writer, there were others who conceived of uses for a high-power, highly directional light beam. Telecommunications engineers envisioned highly directional, line-of-sight communication systems where the information was carried on a light beam. Ophthalmologists needed intense beams of light that could be focused onto small areas at the back of the eyeball to weld detached and torn retinas.

Experience gained in the development of radar during World War II and the continuation of such work at higher microwave frequencies prompted scientists to explore the conditions that were necessary for laser action to be achieved. In the early 1950's, a group at Columbia University headed by Charles H. Townes operated a microwave device that amplified radiation by the stimulated emission process. The device was termed a MASER, an acronym for Microwave Amplification by Stimulated Emission of Radiation. During the remainder of the fifties, the maser principle was employed in many materials. In 1958, Townes and Arthur L. Schawlow published an important paper in which they discussed the extension of maser principles to the optical region of the electromagnetic spectrum. By 1960, a number of groups were investigating systems that might serve as the basis for an optical maser, as it was called by some, or *laser*. Credit for first achieving laser action at optical frequencies is given to T. H. Maiman of the Hughes Research Laboratories. His laser consisted of a pink ruby rod with silvered ends for mirrors inserted in the helical coil of a photographic flashlamp.

Types of Lasers—Edible and Inedible

Within six months of the invention of the ruby laser, laser action was obtained in a mixture of helium and neon gases. Research quickly mushroomed. Many materials were investigated as the active laser medium, including impure crystals, semiconductors, ionized gases, molecular gases, and dye solutions.

Because it is possible to dissolve dyes in many mediums, some of the more exotic types of lasers are dye lasers. A case in point is the edible laser. By dissolving a laser dye in ordinary gelatin (following the directions on the package, according to researchers), a quivering laser medium is produced. A pulse of ultraviolet light is used to excite the medium, and the jellylike material lases. Researchers have even found that the medium will withstand higher pulse repetition rates if it is allowed to quiver than if it is clamped in a rigid enclosure.

The sizes and shapes of lasers are varied. They can be as small as the experimental miniature lasers that are the heart of optical integrated circuits, the future companions of today's semiconductor marvels, or they can fill a room, as do the high-power gaseous lasers that are contorted to achieve a maximum amount of active laser medium in a minimum amount of space. They can be as modest looking as the one-milliwatt educational lasers, or as fearsome as the pulsed, high-power, experimental lasers.

The Present and Future of Lasers

The laser is presently being used in a variety of materials-processing applications. It has become a part of many specialized communication systems, and it is the keystone of modern optical data processing and holographic technology. We will cover these applications in considerable detail in later chapters. The laser is also finding its way into an ever-growing number of other fields: long-distance measurement, construction and airframe alignment, optical spectroscopy, wind velocity measurements, and so on, all of which we cannot cover here. However, these applications are continually being discussed in the trade journals, particularly in those listed among the references for this chapter. Other more specialized books referenced throughout this text discuss some of these fields in detail.

In the future, we can expect lasers of even higher power to be developed. More lasers will have outputs in the ultraviolet end of the spectrum, and possibly in the x-ray region. More efficient lasers may be invented, and lower-cost lasers will be produced. Efforts will continue to modify the output of existing lasers, resulting in new applications. For the present, some basic facts are needed for an understanding of the laser and why it is so valuable to modern technology.

A NOTE ON LASER SAFETY

Although the output of a laser can be a thing of beauty, one must always maintain caution while using these devices. Even the lowly helium-neon laser, used in many laser demonstrations, must be treated with respect and care. Never look into the beam of any laser. Reflections from polished surfaces are as dangerous as the raw beam itself. Do not point a laser at anyone. Always be aware of the beam path and be aware of others when adjusting the path. Do not operate a laser without proper electrical shielding — operating voltages are generally lethal.

The Bureau of Radiological Health of the United States Food and Drug Administration has classified lasers operating in the visible part of the spectrum into four categories, each with specific warning labels and safety requirements. Lasers with outputs in the invisible part of the spectrum must be operated with special care. Copies of Bureau of Radiological Health regulations for laser use should be obtained by anyone using a laser. The sources are given in the references at the end of Chapter 1.

REFERENCES

1.1 M. Ross (ed.) (1972), *Laser Applications, Vol. 1.* New York: Academic Press. Besides the applications covered in this text, the use of the laser as a gyroscope and in ranging (measuring over long distances) and geodesy are discussed.

1.2 S. S. Charschan (ed.) (1972), *Lasers in Industry.* New York: Van Nostrand–Reinhold. Probably the best applications book available to date. Chapter 1 covers much the same material as the first five chapters of this text, but in less detail.

The magazines below report current laser applications. Some may be carried by libraries; others are available only through individuals with complimentary subscriptions. They are indispensible to someone trying to keep up with current applications and products. There is a good deal of overlap in the material covered by all of them.

1.3 *Laser Focus,* Advanced Technology Publications, Newton, Massachusetts. This magazine offers the most comprehensive treatment of laser news. The applications section is particularly good. The magazine is a controlled circulation publication—complimentary subscriptions are available to persons working in laser-related fields.

1.4 *Electro-Optical Systems Design,* Kiver Publications, Inc., Chicago, Illinois. Covers the optical field in general. Large amount of material on laser applications, but also includes optical data processing. Controlled circulation publication exclusively.

1.5 *Optical Spectra,* The Optical Publishing Company, Pittsfield, Massachusetts. Somewhat more emphasis on spectroscopy than the above publications. Another controlled circulation magazine.

The following references relate to laser safety.

1.6 H. Weichel, W. A. D. Danne, and L. S. Pedrotti (1974), "Laser Safety in the Laboratory," *American Journal of Physics* **42**, 1006–1013. An excellent discussion of potential hazards associated with lasers.

1.7 *The ANSI Laser Safety Guide,* ANSI; Z136.1–1976. Published by the American National Standards Institute, 1430 Broadway, New York, NY 10018. Lists maximum permissible exposure levels and safety measures to be employed in using lasers. ($9.00)

1.8 Copies of the federal regulations on lasers may be obtained by writing to the Director (HFX–400), Division of Compliance, Bureau of Radiological Health, 5600 Fishers Lane, Rockville, MD   20852, and asking for HEW (FDA) Pub. 76–803.

Laser Light

The great impact of the laser as an industrial and research tool is the consequence of the extraordinary properties of laser light—extraordinary in the sense that they are present to a far greater degree in the output of a laser than in the light from any other source, natural or man-made. The most striking properties of laser light are its extreme *brightness* and *directionality*. Light from even the most modest of lasers dazzles the eye; the threadlike beam seems never to grow in size. Another unusual characteristic soon evident to even the casual observer is the speckled or scintillating appearance of laser light reflected from a rough surface. This *laser speckle* is a consequence of the extremely high degree of *coherence* of the light from a laser. With proper design, lasers may also exhibit extreme *monochromaticity* and a high degree of *polarization*. Some properties of laser light, such as brightness and directionality, are easily understood, at least qualitatively, since they are evident to any observer. The properties of monochromaticity and coherence offer more difficulty, since one must rely on an instrument other than the eye to evaluate them.

2.1 LIGHT WAVES

To understand the properties of laser light and to understand the laser itself with its many applications, we must first understand the wave nature of light.

The light by which you read these words is wavelike in nature. It is characterized by a combination of time-varying electric and magnetic fields propagating through space. The frequency at which these fields oscillate, v, and their wavelength in a vacuum, λ, are related by

$$\lambda v = c \tag{2.1}$$

where c is the speed of light in a vacuum, approximately 3×10^8 m/sec. In any other medium, $\lambda v = c/n = v$, where n is the refractive index of the medium and v is the velocity of light in the medium. Table 2.1 lists the various units common to the measurement of optical frequencies and wavelengths.

Table 2.1 Optical units.

| **Wavelength** (λ) | | | |
Unit	Designation	Value in meters	Range of common use
Micrometer	μm	10^{-6} m	Infrared
Nanometer	nm	10^{-9} m	Visible and UV
Angstrom*	Å	10^{-10} m	Visible and UV
Frequency (v) is expressed in hertz (Hz). (1 Hz = 1 cps)			
Unit	Designation	Value in hertz	Range of common use
Megahertz	MHz	10^6 Hz	Radio
Gigahertz	GHz	10^9 Hz	Microwaves, infared
Terahertz	THz	10^{12} Hz	Infrared, visible

* The *angstrom* (Å) is still in use in the optical literature; however, by international agreement, quantities that in the past have been expressed in angstroms are now expressed in *nanometers* (nm).

Visible light is not the only kind of electromagnetic wave. The spectrum of electromagnetic radiation extends from the long-wavelength radio waves to x-rays and gamma rays at the shortest wavelengths, as illustrated in Fig. 2.1. Throughout this text, we shall confine our discussion of the properties of laser light to optical wavelengths, i.e., that part of the spectrum extending from the near infrared, through the visible, to the ultraviolet. The visible region of the spectrum (where

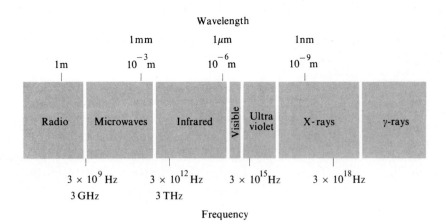

Fig. 2.1 The electromagnetic spectrum.

the human eye is sensitive) encompasses wavelengths from 400 nm (violet) to 700 nm (red). Below 400 nm and extending to 10 nm are the ultraviolet wavelengths; above 700 nm out to about 20 μm are the near-infrared wavelengths.

A discussion of the properties of light waves is greatly aided by a mathematical representation of a wave. In the late nineteenth century, James Clerk Maxwell showed that light waves in free space could be represented by a mathematical expression describing either the electric field or the magnetic field of the wave. Both are not necessary, as their behavior is complementary. Perhaps the simplest electromagnetic wave we can examine is the *monochromatic plane wave*, which is a sinusoidal wave of infinite extent that propagates in a single direction. If we choose the positive z-axis as the direction of wave propagation, the variation of the electric field of the wave in time and space, $E(x,y,z,t)$, is described mathematically by the expression

$$E(x,y,z,t) = A \cos\left[2\pi\left(vt - \frac{z}{\lambda}\right) + \varphi\right] \qquad (2.2)$$

where E is the value of the electric field at time t and at spatial coordinates (x,y,z); A is the amplitude of the wave; v is the frequency in hertz; λ is the wavelength (nm, μm, etc.); and φ is the phase constant. The term in brackets, called the *phase* of the wave, varies both as a function of time and as a function of the distance from the origin in the z-direction. By examining this expression at one instant in time and then again at one point in space, we can obtain a better understanding of the characteristics of the wave. In Fig. 2.2 we have plotted $E(x,y,z,0)$ to show how E varies along the z-axis at the particular instant $t = 0$, assuming that $\varphi = 0$. The corresponding mathematical description, from Eq. (2.2), is

$$E(x,y,z,0) = A \cos\frac{2\pi z}{\lambda}. \qquad (2.3)$$

In the plane $z = 0$, and in all parallel planes an integer number of wavelengths from this plane, the electric field is a maximum. In intermediate planes, the field

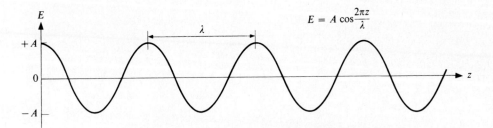

Fig. 2.2 Amplitude of an electromagnetic wave as a function of distance at one instant of time ($t = 0$). The separation between two maxima is the wavelength λ of the wave.

will be smaller in magnitude or negative, depending on the argument of the cosine. For example, any plane where $(2\pi z)/\lambda$ equals an odd multiple of $180°$, $\cos (2\pi z)/\lambda$ is equal to -1, and the electric field has its maximum negative value. The quantity $1/\lambda$ is appropriately referred to as the *spatial frequency* (the number of periods per unit distance) of the wave. We can simplify Eq. (2.3) by substituting the symbol k for $2\pi/\lambda$; k is simply the angular spatial frequency measured in radians per unit distance. With this substitution, the spatial dependence of the light wave at $t = 0$ becomes

$$E(x,y,z,0) = A \cos kz, \qquad k = \frac{2\pi}{\lambda}. \tag{2.4}$$

If we now examine the wave as a function of time as it passes the plane $z = 0$, as shown in Fig. 2.3, E is given by

$$E(x,y,0,t) = A \cos 2\pi vt. \tag{2.5}$$

The number of oscillations of the wave per second at a fixed location equals the frequency, v. Another convenient parameter is $\omega = 2\pi v$, which is the *angular temporal frequency* in radians per second. Expressed in terms of ω, the electric field amplitude of the wave at the plane $z = 0$ becomes

$$E(x,y,0,t) = A \cos \omega t, \qquad \omega = 2\pi v. \tag{2.6}$$

Allowing both t and z to vary again and using the radian measures of temporal and spatial frequency, we can describe the wave in the compact form

$$E(x,y,z,t) = A \cos (\omega t - kz). \tag{2.7}$$

A more general description includes the arbitrary phase constant in the argument,

$$E(x,y,z,t) = A \cos (\omega t - kz + \varphi) \tag{2.8}$$

which shifts the locations of the wave maxima and minima in time and space.

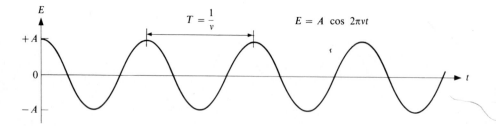

Fig. 2.3 Amplitude of an electromagnetic wave as a function of time at a point in space ($z = 0$). The period T, the time for one cycle, is the reciprocal of the frequency v of the wave.

As time passes, the surfaces of constant phase move through space. These surfaces, or *wavefronts*, are planes in this simple case. Although E is written as a function of x and y as well as of z and t in Eq. (2.8), the only spatial dependence is in the z-direction. Thus the expression describes *plane* waves of infinite extent traveling in the z-direction.

At first, the notion of an infinite plane wave might be difficult to comprehend, it being a mathematical abstraction. In time, however, one comes to recognize the plane wave as the simplest expression we have for a propagating light wave. The simple cosine expression and the idea of an electric field having the same value across large surfaces often serves as a good approximation to real waves.

We can generalize our mathematical description to include a plane wave traveling in an arbitrary direction. Such a wave is characterized by a *propagation vector*, or *wave vector*, $\mathbf{k}$. Propagation vector $\mathbf{k}$ has a length equal to $2\pi/\lambda$, namely

$$k = |\mathbf{k}| = \sqrt{\mathbf{k} \cdot \mathbf{k}} = \frac{2\pi}{\lambda}. \tag{2.9}$$

The direction of $\mathbf{k}$ is the same as the direction of propagation of the plane wave. Under these circumstances, the expression for $E(x,y,z,t)$ becomes

$$E(x,y,z,t) = A \cos(\omega t - \mathbf{k} \cdot \mathbf{r} + \varphi) \tag{2.10}$$

where $\mathbf{r}$ is the vector from the origin to the point (x,y,z). The product $\mathbf{k} \cdot \mathbf{r}$ can be expressed in terms of the vector components of $\mathbf{k}$ and $\mathbf{r}$. For example, if the wave-propagation vector lies in the x–z plane making an angle θ with the positive z-axis, as shown in Fig. 2.4, then

$$\mathbf{k} = k_x\hat{e}_x + k_y\hat{e}_y + k_z\hat{e}_z = k\hat{e}_x \sin\theta + k\hat{e}_z \cos\theta \tag{2.11}$$

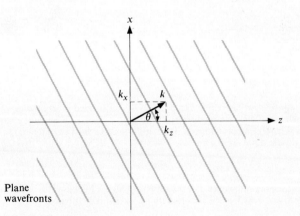

Fig. 2.4 Plane wave with propagation vector $\mathbf{k}$ in the x–z plane (the y-axis is perpendicular to the page). The components of the propagation vector, $k_x = |\mathbf{k}| \sin\theta$ and $k_z = |\mathbf{k}| \cos\theta$, are shown.

and

$$E(x,y,z,t) = A \cos (\omega t - kx \sin \theta - kz \cos \theta + \varphi). \qquad (2.12)$$

The corresponding expressions for waves traveling in other directions are found in the same way.

Of equal importance in understanding the wave nature of light, however, is the concept of a *spherical wave*. Imagine a source of light so small that it is a point of light: a *point source* (see box). Light waves generated by this tiny source radiate out uniformly in all directions; the wavefronts are a series of expanding concentric shells, as illustrated in cross section in Fig. 2.5. The wave maxima of these spherical waves are separated by a distance λ; this time, however, the separation is in the radial direction as measured from the source. If we assume that the source is at the origin of our coordinate system, the mathematical description for the expanding waves is given by

$$E(x,y,z,t) = \frac{B}{r} \cos \left[2\pi \left(vt - \frac{r}{\lambda} \right) \right] = \frac{B}{r} \cos (\omega t - kr) \qquad (2.13)$$

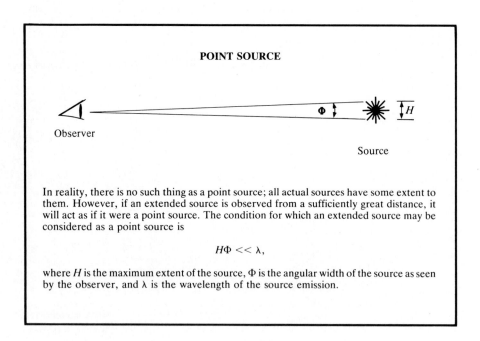

POINT SOURCE

Observer

Source

In reality, there is no such thing as a point source; all actual sources have some extent to them. However, if an extended source is observed from a sufficiently great distance, it will act as if it were a point source. The condition for which an extended source may be considered as a point source is

$$H\Phi \ll \lambda,$$

where H is the maximum extent of the source, Φ is the angular width of the source as seen by the observer, and λ is the wavelength of the source emission.

Fig. 2.5 Cross section of a spherical wave through the point source (S). The lines are loci of maximum positive electric field amplitude. Note that far away from the point source, a spherical wave approximates a plane wave (box with dashed lines).

where B is the maximum amplitude at a unit distance from the source and where $r = (x^2 + y^2 + z^2)^{1/2}$. The appearance of the factor r in the denominator accounts for the $1/r$ decrease in wave amplitude as the wave radius gets larger and larger. A consequence of our idealization is that our expression for E is infinite at $r = 0$, which is a physical impossibility. Although no light source is truly concentrated at a single point, for real sources sufficiently small and r sufficiently large, Eq. (2.13) is an adequate representation of the wave. Note that far away from the source, as in the boxed region of Fig. 2.5, the radii of the spherical waves are so large ($r \gg \lambda$) that these waves are well approximated by plane waves over a limited extent.

If such waves overlap in the same region of space, one simply adds the electric fields of all the individual waves at a point to find the resultant field at that point in space and time. This additivity of electric fields is referred to as the *principle of superposition*.

2.2 MONOCHROMATICITY

The waves we have just discussed are monochromatic; that is, they describe the electric field for light of a single frequency, namely, a single color (Greek: *monos* = single, *chroma* = color). From the standpoint of the laboratory researcher, perhaps the most important property of laser light is its extraordinary monochromaticity. While it is easy to write down an expression for the output of a theoretical monochromatic laser source, it is impossible to determine how close to a single color a laser output may be without the aid of some instrument. What may seem to be a pure color to the unaided eye is in many cases a mixture of several colors. If a prism or diffraction grating is used to disperse the light into its spectrum, one sees that the single-colored nature of various light sources is a matter of degree.

No light source, the laser included, is capable of producing absolutely mono-chromatic light; we can only make better and better approximations to the ideal. We might begin with a white-light source that produces light of all colors of the spectrum, and filter it with a gelatin filter or a piece of colored glass. The mono-chromaticity of the filtered light is now as good as the filter. If the radiation from a gas discharge source, such as a neon sign or a sodium or mercury vapor lamp, is directed through a narrow slit and then a prism, a series of lines of different colors is seen on a screen. The origin of these discrete lines is discussed in Chapter 3. For the present, we need know only that such color lines exist. To speak more quantitatively of the degree of monochromaticity of light from some source, we characterize the spread in frequency of a line by Δv, the *linewidth* of the source, or by the corresponding spread in wavelength, $\Delta \lambda$. (The differences are connected via Eq. (2.1). For small $\Delta \lambda$, one can show that $\Delta v = -(c/\lambda^2) \Delta \lambda$ and $\Delta \lambda = -(c/v^2) \Delta v$.) Depending upon the light source and the level of excitation, the emitted light can consist of color bands that range from the broad (in the case of a white-light source, $\Delta \lambda \cong 300$ nm) to the narrow (for gas discharge lines, $\Delta \lambda \cong 0.01$ nm).* By isolating one of these narrow lines with a suitable filter, we can now achieve monochromaticity as good as the width of a single emission line. If the gas can be made to undergo laser action, this particular line is replaced by a series of even narrower lines; by suppressing all but one of these narrow lines by a tech-nique known as single moding, the most nearly monochromatic source of light known to man can be obtained. But even this exceedingly narrow line contains a small spread of different frequencies. If the light consisted of radiation oscillating at a *single* frequency, the line would be infinitely narrow ($\Delta v = 0$) and the light would be absolutely monochromatic. Absolute monochromaticity is an un-attainable goal that we can only approach by refining our light sources. The single-mode laser, which produces light with the highest degree of monochro-maticity yet attained, falls short of the ideal.

Prior to the development of the laser, high-output power levels of some atomic emission lines were obtained by high-power excitation of a discharge tube. By using rugged filters to absorb all the unwanted radiation, an output approaching that of the laser in monochromaticity and power could be obtained. A laser, however, requires only a modest amount of input power to produce the same effect. The combination of a selective input of excitation and the highly reflective mirrors used in a laser are responsible for intense output from the lasing lines and virtually no output from other lines in the discharge spectrum. In later chapters, we will discuss specific lasers and methods used in obtaining a high degree of monochromaticity.

---

* The terms "broadband" and "narrowband" are often used in describing radiation sources with large or small fractional linewidths, where the fractional linewidth is given by $\Delta v/v$. The terms "bandwidth" and "linewidth" are often used interchangeably.

2.3 DIRECTIONALITY

Perhaps the most arresting property of laser light is its directionality. From the sight of the long, thin output beam, even the most casual observer is aware that the laser is an uncommon source of light. Indeed, the lowest power gas laser made can provide a spectacular light show with the aid of a few mirrors and a little chalk dust or smoke to scatter the light. Even if one were to attempt to collimate the output of a high-power arc lamp, one could not obtain anything approaching the highly collimated output of a small laser.

A handwaving explanation of the source of this high degree of directionality will serve as an introduction to a more realistic explanation of directionality. It might come in handy when your Great Aunt Ethelrude asks her niece or nephew, the Laser Expert, why those lasers give such a tiny beam of light. The light in a laser is contained between two highly reflective mirrors. As a first approximation, one can think of these mirrors as collimating apertures. The mirrors have such a high reflectivity that a wave is reflected many times with only a small portion of the wave being transmitted by the mirrors. The multiple reflections increase the distance the light travels, while confining it within a very small region between the mirrors. Thus, because of the long distance traveled, the curvature of the waves is very small and the light waves emerging from the laser are nearly planar. If we consider a single point source located between the two mirrors, we can get some idea of the collimating property of multiple reflections using only geometry. Each time a light wave reflects off a mirror, the distance between it and the point source that generated it increases. Since the output mirror now transmits a smaller region of that wavefront after each reflection, it serves as a collimating aperture for the wave. If you sit in a barber or beautician's chair between mirrors on opposite walls and see that corridor with you and all your reflections getting a haircut, you have a graphic example of this collimation property. In many lasers, the number of reflections is over 50 and as high as several hundred (this number can be calculated from the mirror transmittance). The beam transmitted from the tube looks as though it comes from a point some 100 cavity lengths distant with 100 apertures collimating the beam.

The directionality of the laser beam is expressed in terms of the *full angle beam divergence*, which is twice the angle that the outer edge of the beam makes with the center of the beam (Fig. 2.6). The divergence tells us how rapidly the beam spreads when it is emitted from the laser. Although the divergence angle can be given in fractions of degrees, minutes, or seconds, it is customary to specify the beam divergence in radians, where 2π radians equals 360°. For a typical small laser, the beam divergence is about 1 milliradian (10^{-3} radians, designated 1 mrad). Using the small angle approximation ($\tan(10^{-3}$ rad$) \doteq 10^{-3}$ rad), one can easily show that this typical laser beam increases in size about 1 mm for every meter of beam travel.

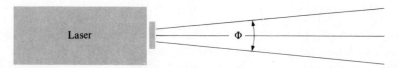

Fig. 2.6 Full-angle beam divergence of a laser is twice the angle between the edges of a laser beam and the beam center. Usually the edge of the beam is defined as the point where the strength of the beam has dropped to 37 percent (1/e) of the value at the center of the beam.

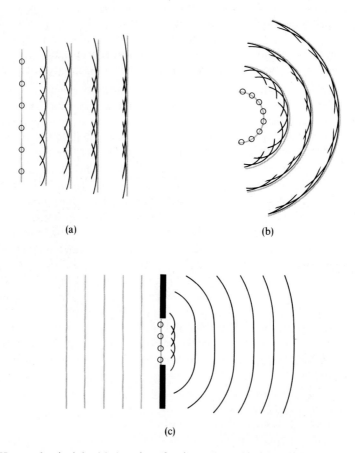

(a)   (b)

(c)

Fig. 2.7 Huygens' principle. (a) A series of point sources (designated by open circles) on a plane wavefront emit secondary wavelets (circular sections to the right). The wavefront at a later time is the resultant of the wavelets, in this case a plane wave. (b) A similar example for a spherical wavefront. (c) An illustration of diffraction through an aperture.

Huygens' Principle and Diffraction

The divergence of a laser beam is a consequence of the natural spreading in space of the beam as the light waves move through space. One can predict how the beam spreads or changes as it propagates by invoking *Huygens' principle*, which says that *the propagation in time and space of a wavefront can be found by replacing the wavefront with a series of equally spaced secondary point sources, all emitting in phase, located on the wavefront.* Using the principle of superposition, we sum the resultant spherical wavelets emitted from these point sources. The summation is equivalent to the wavefront at a later time. In Fig. 2.7, we give several examples. Starting with a plane wavefront (Fig. 2.7a), the wavelets produce another plane wavefront some distance from the original. Similarly for the spherical wave (Fig. 2.7b), the application of Huygens' principle results in a spherical wave of larger radius. In Fig. 2.7(c), the technique is applied to predicting the wavefield that results when an opaque screen containing an aperture is illuminated by a plane wave. Note that the light spreads into regions commonly thought of as shadow regions.

This spreading of the wavefront is called *diffraction*, a natural consequence of the wave nature of light. Of particular importance is the example shown in Fig. 2.8, where a small circular hole of diameter D is illuminated with plane waves

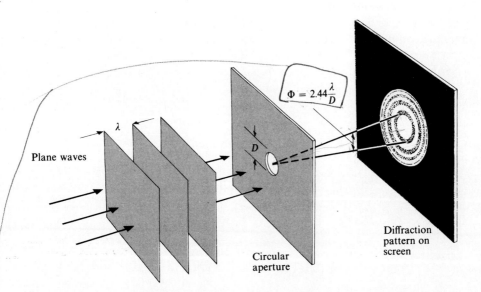

Fig. 2.8 Diffraction of light by a circular aperture. The light passing through the small aperture spreads to form a concentric ring pattern. A measure of the amount of diffraction is the angular diameter Φ of the first dark ring.

of wavelength λ. The light pattern appearing on the screen behind the hole, called a *diffraction pattern*, consists of a central bright spot surrounded by concentric dark and light rings. This diffraction pattern can be calculated, using Huygens' principle, by summing together the fields produced by equally spaced, equal amplitude, spherical wave sources in the aperture. Because 84 percent of the light passing through the aperture is contained in the central bright spot, a measure of the amount of diffraction is given by the angle Φ, subtended by the diameter of the first dark ring. This angle is given by the relationship

$$\sin \Phi = 2.44 \frac{\lambda}{D} \tag{2.14}$$

where λ is the wavelength of the radiation, Φ is the angular diameter of the first dark ring of the diffraction pattern, and D is the diameter of the hole. Note that $\sin \Phi$, and therefore Φ, decreases with increasing hole diameter; that is, the smaller the hole, the greater the spreading of the waves. For D large compared to λ, the diffraction angle is small, and the sine of the angle is approximately equal to the angle in radians:

$$\sin \Phi \cong \Phi = 2.44 \frac{\lambda}{D}. \tag{2.15}$$

In many applications, the output beam from a laser is focused, expanded, or otherwise operated on by lenses, prisms, stops, or apertures. In each case, the edges of a lens, mirror, or other optical component may serve as an effective aperture, limiting the extent of the beam and thereby introducing additional diffraction spreading. The divergence of the laser beam itself, however, is generally determined not by an aperture, effective or otherwise, but by the intrinsic size of the beam within the laser's optical cavity.

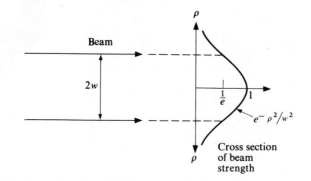

Fig. 2.9 Gaussian beam. The Gaussian beam strength has a smooth exponential dependence as a function of distance from the center of the beam.

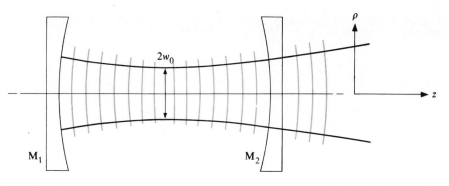

Fig. 2.10 Laser cavity consisting of two curved mirrors, M_1 and M_2. Curvature of the mirrors contains the light within the cavity and causes the Gaussian beam to narrow down to a radius w_0, which is the minimum spot size. The radius of the beam in the waist region determines the amount of divergence of the laser beam. This figure is symmetrical about the z-axis; ρ is the cylindrical radial coordinate.

To confine the laser light within the cavity, the laser mirrors have some positive (inward) curvature, which tends to focus the light as it bounces back and forth. In most lasers, the beam is circularly symmetric in cross section, being most intense at the center and falling off as $\exp\{-(\rho/w)^2\}$ with the distance ρ from the center of the beam. Such a beam, called a *Gaussian beam*, is illustrated in Fig. 2.9. The value of ρ for which the beam irradiance (the power per unit area) decreases to $1/e$ of its value at the center is termed the *spot size, w*. In general, w will vary from point to point along the axis of the beam. At one point in the cavity, called the *beam waist*, the Gaussian beam has its minimum spot size, w_0, as shown in Fig. 2.10.

The spot size at the beam waist determines the divergence of the laser beam, since the external beam is really an extension of the internal laser beam beyond the laser mirrors. Far away from the beam waist, the full angle divergence in terms of the spot size is given by*

$$\Phi = \frac{2\lambda}{\pi w_0} = \frac{4}{\pi} \cdot \frac{\lambda}{2w_0} \cong 1.27 \frac{\lambda}{2w_0}. \tag{2.16}$$

This expression resembles the diffraction pattern from an aperture, with $2w_0$ in place of the aperture diameter D. The factor 1.27, however, is approximately half

---

* Huygens' principle, and the more precise formulations provided by diffraction theory, predicts that the beam remains Gaussian in profile, only spreading as it propagates.

as large as the corresponding factor for a hard-edge aperture (Eq. 2.14). For example, a visible ($\lambda = 500$ nm) laser with Gaussian output and a 1/4 mm beam waist has a full angle divergence of

$$\Phi = \frac{1.27 \times 5 \times 10^{-7} \text{ m}}{0.5 \times 10^{-3} \text{ m}} = 1.27 \times 10^{-3} \text{ rad} = 1.27 \text{ mrad}.$$

In comparison to an ordinary incandescent light source or discharge tube, the laser's directionality is far superior. The difference between a nonlasing source and a laser lies in the method by which the wavefronts are emitted from the sources. For example, a laser medium may be a gas discharge between two mirrors. The focal properties of the mirrors confine the strong laser light to a region around the optic axis of the cavity and causes the wavefronts to be planar in the region of the beam waist. In an ordinary gas discharge, there is no control over the wavefront. They are highly irregular and, therefore, are subject to much greater divergence than most laser sources.

2.4 BRIGHTNESS

Lasers are bright and intense light sources. A one-milliwatt helium-neon laser is, in fact, brighter than the sun! To demonstrate this, we must have some basis for making a comparison of brightness, a quantity that depends on the collimation of the source, the total power emitted, and the response of the human eye. Specifically, we must learn how to measure light.

Radiometry and Photometry

The measurement of electromagnetic radiation is known as *radiometry*. In this discipline, radiation of all wavelengths is treated equally: Radiant energy and power are measured in joules and watts, just as are mechanical and electrical energy and power. A subdivision of radiometry known as *photometry* is restricted to the measurement of *visible* radiation. In this field, the measurement of detected radiometric radiation is weighted according to the sensitivity of the human eye to the color of the light. Yellow light is assigned a higher photometric value than red or blue light because it appears brighter to the human eye. Infrared and ultra-violet light are assigned zero weight because we cannot see these radiations. All the quantities that can be measured by radiometry have counterparts in photometry. To give a feel for the relative brightness of various sources, we must use photometric units when making our comparisons. In most cases, however, we use radiometric units, since they are used routinely in the literature of laser technology and laser safety. We have listed the most important radiometric and photometric quantities, along with their symbols and units, in Table 2.2. Note that the only differences in the terms for energy and power are the prefixes "radiant" for radiometric units and "luminous" for photometric units. The symbols for the quantities are the same except for "e" (for energy) subscripts for radiometric quantities and "v" (for visual) subscripts for photometric quantities.

Table 2.2 Radiometric and photometric quantities.

Radiometry			Photometry		
Term	Symbol	Units	Term	Symbol	Units
Radiant energy	Q_e	Joule	Luminous energy	Q_v	Talbot
Radiant power or flux	Φ_e	Watt (equals 680 lumens at 555 nm)	Luminous power or flux	Φ_v	Lumen
Irradiance	E_e, I	Watts/m²	Illuminance	E_v	Lumens/m²
Radiant intensity	I_e	Watts/sterad	Luminous intensity	I_v	Lumens/sterad
Radiance	L_e	Watts/m²-sterad	Luminance (Brightness)	L_v	Lumens/m²-sterad

1. Any term can be taken per unit frequency spread or wavelength bandwidth. When these quantities are used, the adjective "spectral" is added as a prefix to that term. Either a "v" is added as a subscript to the symbol and the quantity is divided by the frequency spread in units of frequency (Hz), or a "λ" is added as a subscript and the quantity is divided by the bandwidth in units of wavelength (μm, nm, etc.).

2. The area denoted in the terms for radiance and luminance is the area of the source. The area denoted in the irradiance and illuminance is the area of the illuminated surface. Light radiating from the source area irradiates a surface. (The corresponding sentence using photometric terms is awkward, since there is no gerund form for the noun "luminance.")

The link between radiometric and photometric units is the *standard luminosity curve*. This curve, shown in Fig. 2.11, represents the response of the average human eye to light of various wavelengths. The curve peaks in the yellow and falls away on both the red and the violet ends of the spectrum. The single point of contact between radiometry and photometry is the defining statement that 680 lumens (a photometric unit) is equal to 1 watt (a radiometric unit) at 555 nm (bright yellow), the peak of the standard luminosity curve. At any other wavelength, this conversion is scaled by the value of the standard luminosity curve at the wavelength. Outside the visible wavelength region, even the strongest source has zero value in photometric units. Thus the output of infrared and ultraviolet lasers must be measured in radiometric units. In this section, we assume that we are dealing with sources having a continuous output. It should be noted, however, that many lasers are pulsed. In these cases, the quantity of interest may not be average radiant power (watts) or irradiance (watts/m²), but the total radiant energy in a single pulse (joules) or the radiant energy density (joules/m³).

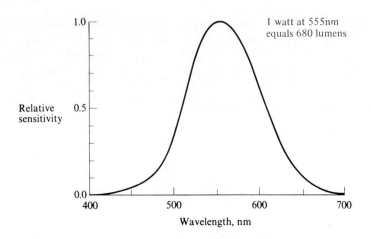

Fig. 2.11 The standard luminosity curve indicates the response of the average human eye to colors. The curve was generated by asking people to compare the brightness of two lights of different colors and then recording the relative power needed to obtain lights of equal brightness as determined by the observers. The curve shown represents the response of the eye in a light-adapted state (photopic vision). There is another curve for the eye in a dark-adapted state (scotopic vision) that peaks as a shorter wavelength.

Several terms require additional comment. In the strict sense, the term "intensity" applies only to the amount of power radiated per unit solid angle (for a review of solid angle, see box, p. 21): watts/steradian (radiometric); lumens/steradian (photometric). The quantity that describes the power incident on a unit area, popularly known as the "intensity" of the light, is properly called the *irradiance* in radiometry (watts/m^2) and the *illuminance* in photometry (lumens/m^2). Since we want to compare the brightnesses of sources, the most useful quantities are *radiance* (watts/m^2-sterad), the radiometric term, and *luminance* (lumens/m^2-sterad). This latter term, which is commonly called the *brightness*, is the luminous power per unit area of the source per unit solid angle into which the source is radiating.

A Laser Is Brighter Than the Sun!

The sun has a brightness of 1.5×10^5 lumens/cm^2-sterad. How is this powerful source that radiates in all directions to be contrasted to the brightness of a one-milliwatt helium-neon laser, which has a highly directional beam and narrow bandwidth? For small lasers, such as the one-milliwatt laser, the beam divergence is about 10^{-3} rad. The corresponding solid angle of the beam is about 10^{-6} sterad (see box). The beam diameter as it leaves the output mirror of the laser is about 1 mm. Using the illuminated area of the mirror as an approximation to the source area, we get an area of 7.85×10^{-3} cm^2 ($A = \pi r^2$, where $r = 0.5$ mm). The output power of the laser, one milliwatt, is given in radiometric units. Converting to photometric units, we multiply the power (10^{-3} watts) by the conversion factor at 555 nm, 680 lumens/watt, and by the relative luminosity at 633 nm (the wavelength of the laser output) to get a luminous flux of 0.16 lumens for a one-milliwatt laser output at 633 nm. Dividing this value by the source area and

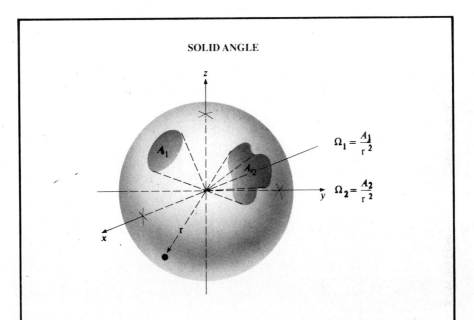

SOLID ANGLE

$$\Omega_1 = \frac{A_1}{r^2}$$

$$\Omega_2 = \frac{A_2}{r^2}$$

A plane angle, expressed in radians, is given by the ratio of the portion of the circle sub-tending the angle to the radius of the circle. The solid angle, in direct analogy, is the area of the portion of the sphere subtending the solid angle divided by the square of the sphere's radius, as shown in the accompanying figure. The unit of solid angle is the steradian (from *stereo*, meaning solid, and radian). An entire sphere, for example, has a solid angle given by $4\pi r^2$ (the surface area of the sphere) divided by r^2 (the square of the radius), or 4π steradians. Note that the solid angle being characterized can have an irregular cross section, as has Ω_2 in the figure; it need not be circular or square.

The solid angle is a dimensionless quantity, since it is an angular measure independent of the size of any defining sphere. As an example, consider that a laser with a 1 milliradian full angle beam divergence illuminates an area π mm^2 on a screen placed 2 meters from the laser. The solid angle subtended by the beam is $\pi(10^{-3})^2$m^2/4m^2, or approximately 10^{-6} steradians.

the solid angle of the beam, we obtain a value for the brightness:

$$L_v = 2.04 \times 10^7 \text{ lumens/cm}^2\text{-sterad}$$

which is a hundred times brighter than the sun!

This does not tell the whole story, however. After all, the visible output from the sun is spread across the entire visible portion of the spectrum, while the laser output is confined to a wavelength spread of a fraction of a nanometer. Any of the terms we have discussed (Table 2.2) can be further specialized by calculating the quantity for a unit spectral bandwidth, as we noted in footnote 1 to the table. For visible lasers, the most important photometric quantity is the spectral brightness or spectral luminance, $L_{v\lambda}$, given in lumens/cm^2-sterad-nm.

The average spectral brightness of the sun is its luminance divided by the bandwidth of the sun's visible output, which we take to be the entire visible wavelength region 400 to 700 nm, or 300 nm. Thus the spectral brightness of the sun is

$$L_{v\lambda} = \frac{1.5 \times 10^5 \text{ lumens/cm}^2\text{-sterad}}{300 \text{ nm}}$$

$$= 500 \text{ lumens/cm}^2\text{-sterad-nm.}$$

The one-milliwatt helium-neon laser has a bandwidth of approximately 0.2 nm, giving a spectral brightness of

$$L_{v\lambda} = \frac{2 \times 10^7 \text{ lumens/cm}^2\text{-sterad}}{0.2 \text{ nm}}$$

$$= 10^8 \text{ lumens/cm}^2\text{-sterad-nm.}$$

No one-milliwatt laser will ever light up a room or illuminate a beautiful landscape as the sun does. There is no better source than the sun for lighting large areas, as any photographer will testify. But as a *compact, collimated, spectrally pure* light source, the laser stands alone.

This enormous spectral brightness presents a problem to those who use lasers. In the past, our experience with bright sources has conditioned us to avoid looking at them. Usually we have some warning of extreme brightness from the level of illumination in an area surrounding the source. Except when a sudden reflection of the sun on a shiny surface bounces light into our eyes, we do not look directly at the noonday sun (save for those foolhardy viewers of solar eclipses). But the extreme directionality of the laser output prevents an observer from seeing the path of the beam unless there is considerable scattering of the light by dust or vapor. If one is not careful, an unguarded beam or an errant reflection can cause temporary loss of sight. In some cases, vision may be permanently impaired. Even when the divergence of a beam has been increased intentionally, it is still possible for sufficient light to be focused on the retina by the lens of the eye to cause damage. One must take into account such things as the laser output power, beam divergence, reflectivity of illuminated surfaces, wavelength of output, and duration of output before operating a laser in a new environment. By learning the current laser safety standards and understanding the radiometric quantities presented here, one can proceed to calculate permissible levels of radiation in a laboratory setting or for an application.

2.5 LIGHT INTERFERENCE: THE LASER SPECKLE PATTERN

When laser light is reflected from a rough surface, one usually notices a peculiar texture to the light—a sparkling or speckled pattern quite unlike any other light patterns. When the observer moves his head, the pattern also moves. This pattern, seen on surfaces illuminated by laser light, is referred to as *laser speckle*. Although a nuisance in some laser applications, laser speckle is a manifestation of another very important property of laser light, coherence, which we discuss in detail in the next section.

Laser speckle can be understood in terms of the phenomenon of *light interference*. We begin our explanation with a simple experiment first performed by Thomas Young* at the beginning of the nineteenth century. Consider an opaque screen containing two pinholes a distance d apart, as illustrated in Fig. 2.12(a). The screen is illuminated by a monochromatic light wave.[†] The two pinholes act as point sources, and consequently the light from the pinholes has the form of two overlapping spherical waves. The wavefronts in a plane containing the pinholes resemble the ripples on a pond when two pebbles are thrown into the water a small distance apart. Employing the principle of superposition, we see that at some points in the water where a crest meets a crest, the ripples are quite high; at other points where a crest meets a trough, the ripples "cancel out."

The light pattern observed on a screen far from the pinholes consists of a series of alternating light and dark parallel bands, or *fringes*. Looking at the figure of the overlapping waves, we see that there are places where the total electric field is twice the value that would be observed with a single pinhole source, and there are other places where the component fields are always 180° out of phase and therefore always cancel. The result is an irradiance pattern that is alternately bright and dark.

---

* Thomas Young (1773–1829), physicist and physician, established the wave theory of light and discovered the principle of light interference.

† We assume monochromatic light because the algebra is simple, and effects using non-monochromatic light are easily explained in terms of the monochromatic case.

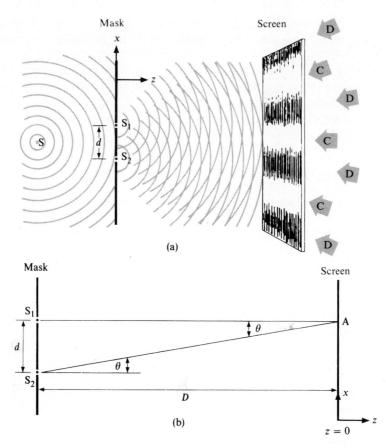

Fig. 2.12 Young's experiment. (a) A point source (S) illuminates two pinholes (S$_1$ and S$_2$), which serve as two secondary point sources. Lines of constructive interference are indicated by arrows ◁ C behind the screen. Lines of destructive interference are indicated by ◁ D. (b) Geometry used to find the interference fringe formula.

    Let us examine analytically the formation of this fringe pattern. For simplicity, we assume that the distance from the pinholes to the screen, D, is much larger than the separation, d, between pinholes. Under these circumstances, the spherical waves incident on the screen can be approximated by two plane waves over a small region. We examine the waves in the region about the point labeled A in Fig. 2.12(b) directly opposite one of the pinholes. About this point, one wave arrives with wavefronts parallel to the screen, the other with wavefronts at an angle θ to the screen, as shown in Fig. 2.13. These two waves, described by Eqs.

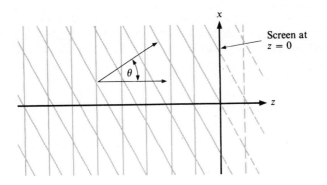

Fig. 2.13 Interfering wavefronts at the screen ($z = 0$). Lines are locations of maximum positive wave amplitude at a particular point in time.

(2.8) and (2.12), have the form

$$E_1(x,y,z,t) = A \cos (\omega t - kz + \varphi_1) \tag{2.17}$$

and

$$E_2(x,y,z,t) = A \cos (\omega t - kx \sin \theta - kz \cos \theta + \varphi_2) \tag{2.18}$$

where we have assumed equal amplitudes. The phase constants, φ_1 and φ_2, depend upon the locations of the pinholes relative to the source and upon their separation. In general, by appropriate choice of coordinate system and pinhole positions, we can cause φ_1 and φ_2 to be zero. It is instructive to carry the arbitrary constants along in our calculations, however, because when we consider non-monochromatic illumination, φ_1 and φ_2 assume additional significance, as we shall discuss later.

Using the principle of superposition, we add the two waves to obtain

$$E_{\text{total}}(x,y,z,t) = A \cos (\omega t - kz + \varphi_1)$$
$$+ A \cos (\omega t - kx \sin \theta - kz \cos \theta + \varphi_2). \tag{2.19}$$

Neither our eyes nor any other light detector responds directly to the rapidly varying* electric field of the waves at the screen, but rather to the time average of the square of the electric field, the irradiance.[†] We must therefore calculate the expression

$$I(x,y,z) = C \frac{1}{T} \int_0^T E^2(x,y,z,t) \, dt \tag{2.20}$$

---

* Roughly 5×10^{14} hertz.

[†] To avoid confusion with the electric field amplitude, E, we will use I instead of E_e to denote irradiance.

in the plane of the screen. The interval T is chosen such that $T \gg 1/\omega$; C is a proportionality constant that relates the square of the electric field (volts^2/m^2) to the irradiance (watts/m^2). Since we are looking in a plane where z is constant (we want to know how the irradiance varies in the x-direction, i.e., along the screen), let us set $z = 0$, with the result

$$E_{\text{total}}(x,y,t) = A \cos(\omega t + \varphi_1) + A \cos(\omega t - kx \sin\theta + \varphi_2) \qquad (2.21)$$

If we use the trigonometric identity $\cos\alpha + \cos\beta = 2\cos[(\alpha+\beta)/2]\cos[(\alpha-\beta)/2]$, we can rewrite the expression for the electric field at the screen in the form

$$E_{\text{total}}(x,y,t) = 2A \cos\frac{1}{2}(2\omega t - kx \sin\theta + \varphi_1 + \varphi_2)$$

$$\times \cos\frac{1}{2}(kx \sin\theta + \varphi_1 - \varphi_2). \qquad (2.22)$$

Now we have a separation of the expression into two factors, one of which is time-independent. The corresponding equation for the irradiance is

$$I(x,y) = \frac{4CA^2}{T} \cos^2\frac{1}{2}(kx \sin\theta + \varphi_1 - \varphi_2)$$

$$\times \int_0^T \cos^2\frac{1}{2}(2\omega t - kx \sin\theta + \varphi_1 + \varphi_2)\, dt. \qquad (2.23)$$

The integral, evaluated for $T \gg 1/\omega$, is equal to $T/2$, yielding

$$I(x,y) = 2CA^2 \cos^2\frac{1}{2}(kx \sin\theta + \Delta\varphi)$$

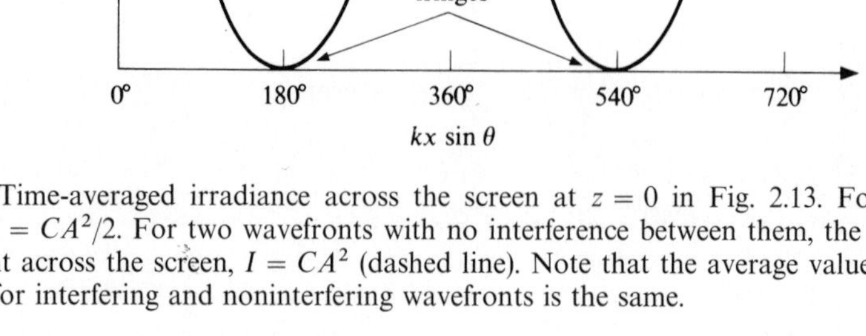

Fig. 2.14 Time-averaged irradiance across the screen at $z = 0$ in Fig. 2.13. For a single pinhole, $I = CA^2/2$. For two wavefronts with no interference between them, the value of I is constant across the screen, $I = CA^2$ (dashed line). Note that the average value of the irradiance for interfering and noninterfering wavefronts is the same.

or

$$I(x,y) = CA^2[1 + \cos(kx \sin \theta + \Delta\varphi)] \qquad (2.24)$$

where $\Delta\varphi = \varphi_1 - \varphi_2$. The irradiance varies cosinusoidally in the x-direction and is plotted in Fig. 2.14. (In obtaining the final expression (Eq. 2.24), we have used the trigonometric identity $\cos^2 \alpha = (1 + \cos 2\alpha)/2$.)

This phenomenon, the generation of alternating bright and dark fringes in an irradiance pattern by the superposition of two waves, is referred to as *interference*. The fringe pattern itself is called an *interference pattern*. These fringes in the irradiance distribution occur because the phase relationship between the two waves is constant with time. At certain points on the screen, the electric field contributions of the two waves are always equal in magnitude but opposite in sign, and the net result at these points is zero electric field, hence zero irradiance. Interference at these points is termed *destructive*. There are other locations on the screen where the wave contributions are equal in magnitude and of the same sign. Here the net electric field is twice the value obtained with a single slit, and the irradiance is quadrupled. Interference at these points is known as *constructive*. If there were no interference between the waves from the two pinholes, the irradiance would be $I = 2(CA^2/2)$, which is twice the irradiance for a single pinhole. The dotted line in Fig. 2.14 is the value of the irradiance if there were no interference. The irradiance for the interfering wavefronts averages to the same value over one fringe. Thus energy is conserved in this interference pattern; it is merely distributed differently.

The concepts of *relative phase* and *pathlength difference* are important to the understanding of interference. If the two waves oscillate at a particular point on the screen as $\cos(\omega t + \psi_1)$ and $\cos(\omega t + \psi_2)$, we say that the waves have a *relative phase* of $[(\omega t + \psi_1) - (\omega t + \psi_2)] = (\psi_1 - \psi_2)$ radians. When the relative phase equals an even-integer multiple of π radians, i.e., when $(\psi_1 - \psi_2) = n\pi$, $n = 0, \pm 2, \pm 4, \pm 6, \ldots$, the waves are said to be *in phase*, and constructive interference occurs. When $(\psi_1 - \psi_2)$ equals an odd-integer multiple of π radians, $(\psi_1 - \psi_2) = n\pi, n = \pm 1, \pm 3, \pm 5, \ldots$, the waves are 180° out of phase, and destructive interference occurs. The relative phase of the two interfering waves at a point on the screen depends upon two factors: (1) the relative phase of the waves at the pinholes and (2) the difference in pathlengths traveled by the two waves in reaching that point. Assume for convenience that the waves at the pinholes oscillate with the same phase; i.e., the relative phase at the pinholes is zero. If the distance to the point from pinhole 1 is smaller by an amount δ than the distance from pinhole 2, then the relative phase is increased by an amount $k\delta = (2\pi\delta)/\lambda$. A pathlength difference of a half wavelength ($\delta = \lambda/2$) therefore results in a phase change of $\Delta\psi = \pi$ (or 180°) compared to $\Delta\psi = 0$ at equal pathlengths ($\delta = 0$). A change in pathlength difference by $\lambda/2$ results in a variation in irradiance from minimum to maximum. Obviously, light interference techniques are excellent methods for measuring small variations in distance.

Interference is the cause of laser speckle. Consider a rough surface illuminated with laser light. Each point on the surface scatters the light as a spherical wave, and thus we can approximate the surface as a collection of a great number of closely packed point sources. Because of the roughness of the surface, the relative phases of these point sources are random. They are, however, constant in time so long as the surface is not moving. When we look at such a surface, each point of light is imaged onto the retina of the eye. The individual point-source images are themselves not points of light, however; the pupil of the eye serves as a circular aperture that diffracts the light waves as they enter the eye. The individual point images are blurred as a consequence and overlap with one another. At any given point on the retina, there are wave contributions from a number of different points on the illuminated surface, each with a random phase. The superposition of these waves can have a very large or a very small net electric field. The result is an interference pattern on the retina that is highly irregular—the speckle pattern. When the head moves, the optical pathlengths from the individual surface points to the retina change, and the speckle pattern itself changes. Because the pathlength changes are coordinated, the speckle pattern usually appears to "move" as the head moves.*

2.6 COHERENCE

Our explanation of laser speckle is thus far incomplete, because we have not explained why such an effect is so easily seen with laser light and not with light from any other source. Laser speckle is a result of interference; interference, in turn, is a manifestation of *coherence*. Normally, our appreciation of coherence—the orderly connection of notes in music, shapes and colors in art, or ideas and events in our lives—is a very different sort from the coherence of light. But even with light one can draw parallels. When a light wave exhibits complete coherence, there is a predictable connection or correlation between the amplitude and phase at any one point on the wave and at any other point; maximum amplitude at a point in space and time *implies* to some degree the amplitudes and phases at all other points on the wave. The highest degree of coherence among light sources is found in lasers. An immediate consequence is the appearance of speckle; of greater significance is the applicability of laser light in such areas as holography and interferometry.

---

* This "movement" of the speckle pattern bears an interesting relationship to the state of correction of the eye. For nearsighted people, the pattern moves against the background opposite to the direction that the head moves. For farsighted people, the pattern moves in the direction of head motion. Unless they intentionally defocus their eyes, individuals with perfectly corrected vision observe no movement at all.

Correlation

The most useful concept to understand when speaking of coherence is *correlation.* Correlation is a matter of degree. We speak, for example, of the degree of correlation between the number of smokers in this country and the incidence of lung cancer. Correlation in time hinges on the predictability of one event or action based on a knowledge of an earlier event. Such correlation often depends on the cause-and-effect relationship between the events, such as the drinking of hemlock and the death of the drinker (a very high degree of correlation!). There also can be correlation in space, a simple example being the little people in Fig. 2.15. If we know the location of one marcher, it is highly probable that we shall find another marcher three feet to the left and three feet behind. Any parade should provide examples of all degrees of coherence, both in space and in time.

Coherent   Incoherent

Fig. 2.15 Coherence and incoherence. With complete coherence, the location of one marcher enables one to predict the locations of all the other marchers.

If light from a source is completely coherent both in space and in time, there is complete correlation between the electric field variations of the light at any one point in space and those at any other point. Having measured once the electric field variations at both points, one can say with complete certainty at any later time what the electric field is doing at the second point simply by measuring the field at the first point. The key lies in the form the electric field amplitude assumes under conditions of complete coherence. It can be shown that only light that is monochromatic can be completely coherent in both time and space. The electric field amplitude then has the form $E(x,y,z,t) = A(x,y,z) \cos [\omega t + \theta(x,y,z)]$, where $\theta(x,y,z)$ is the space-varying part of the phase. Variations in the field at any point in space are always sinusoidal with radian frequency ω. Knowledge of A and θ as functions of position allows us to specify E for all points in space and for any time t, as required of a completely coherent light wave.

Measuring Coherence

As implied by our discussion of correlation, the coherence of light is a matter of degree: Light can be coherent, incoherent, or partially coherent. Our ability to measure coherence is linked to the relationship between coherence and interference. The optical instruments most often used in the measurement of coherence are, in fact, called *interferometers*. Later in this chapter, we shall consider two of these instruments in some detail, because they are useful in a number of laser-related applications. For the present, however, we consider the measurement of coherence from the standpoint of Young's experiment.

In our discussion of that experiment, two pinholes were illuminated and the interference fringe pattern was observed on a screen some distance from the pinholes (Fig. 2.12a). A measure of the fringe contrast, called the *fringe visibility* (V) can be defined as follows:

$$V = \frac{\text{maximum fringe irradiance} - \text{minimum fringe irradiance}}{\text{maximum fringe irradiance} + \text{minimum fringe irradiance}}. \tag{2.25}$$

Note that the fringe visibility must have a value between zero and unity, because irradiance is always a nonnegative quantity.

The fringe visibility can be taken as a direct measure of the degree of coherence the light exhibits at the two pinholes. In order for V to actually equal the coherence of the light at these two points, we must require that equal amounts of light pass the two pinholes. (Consider the effect of blocking one of the pinholes: The fringe pattern completely disappears and $V = 0$, yet the coherence of the light at the two pinholes certainly has not changed.) Specifically, let I_1 and I_2 be the irradiances that result on the screen from each pinhole separately. If $I_1 = I_2$, the fringe visibility V equals the coherence of the light at the two pinholes. (If $I_1 \neq I_2$, then the coherence equals the fringe visibility multiplied by the factor $(I_1 + I_2)/2\sqrt{I_1 I_2}$.) In the interference experiment discussed in the previous section, $I_1 = I_2$, and the fringe irradiance varied as shown in Fig. 2.14. Here the minimum fringe irradiance is zero, and the fringe visibility is thus unity. The coherence of the illuminating wave at the two pinhole locations is, therefore, also unity; i.e., the light is seen to be completely coherent, consistent with our assumption of monochromatic illumination.

Light waves from nonlaser sources exhibit varying degrees of coherence. If the source illuminating the pinholes is compact, approximating a point source, then light waves striking both pinholes exhibit much the same electric-field variations in time, and fringes appear on the screen. If the coherence of the light does not equal unity, the light at the pinholes is said to be *partially coherent*; the fringe visibility is less than unity. If the light waves passing through the two pinholes bear absolutely no relationship to one another, i.e., the electric-field variations are completely uncorrelated, then the fringe visibility will be zero and the light completely incoherent.

For waves exhibiting partial coherence, the characteristics of the coherence can be investigated in detail. If the pinholes are moved about in the plane where they are located, the coherence of the wave in the direction transverse to the propagation direction can be explored. By holding one pinhole fixed and moving the other pinhole about, one can look for any reduction in the fringe visibility. The area of the wavefront over which the pinholes can be moved and the fringes still be seen is known as the *coherence area* of the light wave. This area is a measure of the *spatial* or *transverse* coherence of the wave; it characterizes the spatial variation in coherence across the wavefronts in the direction transverse to the propagation direction.

Besides transverse coherence, a wave possesses some degree of *longitudinal* or *temporal* coherence. This type of coherence refers to the correlation between the wave at one point and the wave at the same point at a later time, that is, to the relationship between the field $E(x_1,y_1,z_1,t_A)$ and the field $E(x_1,y_1,z_1,t_B)$. A variation of Young's experiment can be constructed that explores this relationship. A separation in time in a wavefield corresponds to a distance along the direction of propagation, the z-direction, for example. Since the wave velocity is c, the distance traveled by the wave during the time interval $(t_B - t_A)$ is given by

$$\Delta z = c(t_B - t_A). \tag{2.26}$$

We therefore must modify Young's experiment to look at points separated by Δz along the direction of propagation. This is achieved by an arrangement of beamsplitters (BS) and movable mirrors (M_1 and M_2), as illustrated in Fig. 2.16.

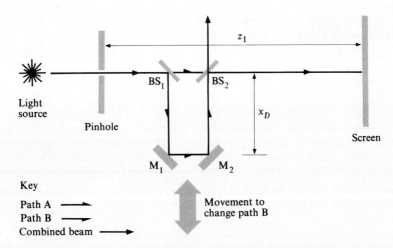

Fig. 2.16 Modified Young's experiment to test temporal coherence. The delay in the reflected beam is accomplished by a movement of the mirrors M_1 and M_2.

Light passes through a single pinhole and is split into two beams by a half-silvered mirror. The light is later recombined at a second half-silvered mirror. One part is incident on the screen after a travel time of z_1/c. The other portion is delayed and reaches the screen after a travel time of $(z_1 + 2x_D)/c$. By varying x_D, we can explore the correlation of the wave for different time delays.

For very small values of x_D,* fringe visibility remains high, and we conclude that the mutual coherence for wave segments closely spaced in time is near unity. As x_D increases, fringe visibility decreases, and ultimately we reach a separation where the fringes disappear completely. This distance we define as the *coherence length* of the radiation, l_c. If the speed of light in the medium is c, then the corresponding time delay, known as the *coherence time* of the source, t_c, is given by

$$t_c = \frac{l_c}{c}. \tag{2.27}$$

There is a direct relationship between the degree of monochromaticity of a light source and the degree of temporal coherence it exhibits. If the coherence length is extremely long, the wave has to be extremely regular for long periods of time. The closer the wave approaches a true sinusoid, the longer the fringe visibility remains near unity as x_D is increased. Thus for nearly monochromatic waves (approaching exact sinusoids), the coherence length is large. In fact, the degree of monochromaticity is related to the coherence length by

$$l_c = \frac{c}{\Delta v} \tag{2.28}$$

where Δv is the linewidth of source radiation. The coherence time of the radiation is then related to Δv through the relation

$$t_c = \frac{1}{\Delta v}. \tag{2.29}$$

For example, a source with a bandwidth of 100 nm centered about $\lambda = 500$ nm ($\Delta v = -c \, \Delta\lambda/\lambda^2$, therefore $\Delta v = 1.2 \times 10^{14}$ Hz) has a coherence length of about 2.5 μm. If the bandwidth were 10^{-3} nm, the coherence length would be 0.25 meter.

The measurement of coherence is usually done not with pinholes in a Young's experiment, but with optical devices called interferometers. We will discuss below two such interferometers that bear a special relationship to the laser and to its applications: the Fabry-Perot interferometer and the Michelson interferometer.

---

* For incandescent sources and the like, x_D must be smaller than approximately 1 μm if fringes are to be observed. For most lasers, the distance can be much larger, in some cases exceeding several kilometers.

The Fabry-Perot Interferometer

A Fabry-Perot interferometer consists of a pair of highly reflective mirrors parallel to each other and whose separation can be varied. The reflectivity of the mirrors is such that when a *single* mirror is illuminated, roughly 95 percent of the incident light is reflected and 5 percent is transmitted, independent of wavelength. For most wavelengths, the mirror *pair* is also highly reflective and only a very small fraction of the incident light is transmitted—consistent with our notion of what should happen when one mirror is placed behind another. For certain wavelengths, however, the mirror pair is highly transmissive, and upwards to 100 percent of the incident light is transmitted. For light traveling perpendicular to the mirrors, this high transmission occurs when d, the distance between the mirrors, equals an integral number of half-wavelengths, i.e., when $d = m(\lambda/2)$, or,

$$\lambda = \lambda_m = \frac{2d}{m}, \qquad m \text{ an integer.} \tag{2.30}$$

We say that the Fabry-Perot mirror *cavity* is *resonant* at these particular wavelengths. It can be shown that when this resonance condition (Eq. 2.30) is satisfied, the waves entering the mirror cavity, the waves leaving the cavity, and the waves bouncing back and forth between the mirrors are *all in phase*. The condition for cavity resonance is thus the condition for constructive interference of the waves within the cavity.

Transmission for the Fabry-Perot is usually expressed in terms of the frequency of the light. Substituting c/v for λ in Eq. (2.30), we obtain

$$v = v_m = \frac{c}{\lambda_m} = m\left(\frac{c}{2d}\right), \qquad m \text{ an integer,} \tag{2.31}$$

as a resonance condition in terms of v. A plot of the transmission of a Fabry-Perot interferometer as a function of frequency resembles a "picket fence" in that there is a series of equally spaced transmission peaks at frequencies given by Eq. (2.31). The separation between neighboring transmitted frequencies is given by

$$\Delta\sigma \equiv v_{m+1} - v_m = (m+1)\frac{c}{2d} - \frac{mc}{2d} = \frac{c}{2d}. \tag{2.32}$$

Note that $\Delta\sigma$ is independent of m. If a single gas discharge emission line, isolated with a filter and incident on the interferometer, has a linewidth that is greater than the separation between transmitted frequencies ($\Delta v > \Delta\sigma$), the transmitted light, plotted as a function of frequency, has a spectrum of several lines, as shown in Fig. 2.17. If the separation between the plates, d, is decreased by a factor of five or so, $\Delta\sigma$ is then large enough so that only one transmission at v_m occurs within the frequency spread of the emission line, as shown in Fig. 2.18(a). If we

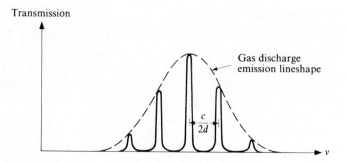

Fig. 2.17 Light of a discharge line transmitted by a Fabry-Perot interferometer of fixed spacing d. Because the separation between resonant frequencies is less than the linewidth, several frequencies within the lineshape are transmitted.

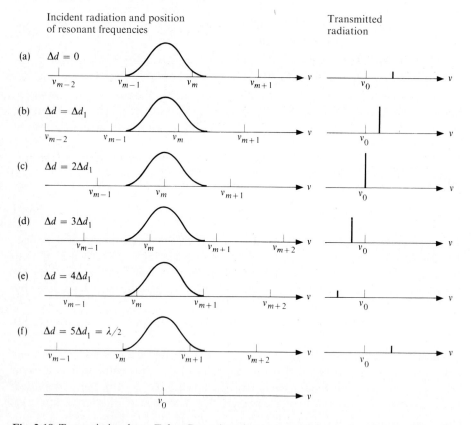

Fig. 2.18 Transmission by a Fabry-Perot interferometer of frequencies of an emission line as a function of small changes in plate separation, Δd, for the case where the free spectral range, $\Delta \sigma$, is greater than the extent of the emission line. Note that when $\Delta d = \lambda_m/2$, the frequencies transmitted are the same as those for $\Delta d = 0$.

change the separation by Δd, where Δd is small compared to a wavelength, the interferometer has a new resonance frequency v_m' given by

$$v_m' = m \frac{c}{2(d + \Delta d)} = m \frac{c}{2d} \left(1 + \frac{\Delta d}{d}\right)^{-1}$$

$$\cong v_m \left(1 - \frac{\Delta d}{d}\right) = v_m - \Delta v \tag{2.33}$$

where $\Delta v \; (= v_m(\Delta d/d))$ is proportional to the change in separation, Δd. (In obtaining this expression, we have used the approximation $(1 + \varepsilon)^{-1} \approx 1 - \varepsilon$ for $|\varepsilon| \ll 1$.) Thus if d is changed by small amounts, various frequencies across the emission line are transmitted, as shown in Fig. 2.18(b–f). In each case, Δd has been changed by the same amount, $\Delta d_1 = \lambda/10$. Notice that in the last figure (Fig. 2.18f), where the cumulative change in the plate separation is a half-wavelength, the transmitted frequencies are the same as in Fig. 2.18(a), where $\Delta d = 0$. Only the value of m is different. Changes of Δd greater than $\lambda/2$ result in the same output as those between zero and $\lambda/2$. The range of frequencies that can be swept through without repetition is $\Delta\sigma = c/2d$. For this reason, $\Delta\sigma$ is called the *free spectral range*.

It is not necessary to change Δd in steps, as has been done here. If it is done in a continuous manner, the output follows the lineshape of the emission line. Thus the Fabry-Perot interferometer can be used either as a spectrometer to determine the lineshape of an emission line or as a tunable filter to transmit a narrow range of frequencies. For example, consider such a filter normally illuminated with green light and adjusted to pass the frequency $v_m = 6 \times 10^{14}$ Hz (corresponding to a wavelength $\lambda = 500$ nm in the green portion of the spectrum). If $d = 10$ cm, then by translating one mirror by 1 nm to either side of its original position, one can sweep the mth resonance frequency through a range of 6×10^6 Hz, or 6 MHz, to either side of the 500 nm line.

The Fabry-Perot interferometer is most useful for examining the frequency content of a laser. As shown in Fig. 2.19, the transmitted beam is focused by a lens onto a screen which contains a small pinhole at point P_1. Only light that is incident parallel to the optic axis of the lens and that satisfies the resonance condition (Eq. 2.31) is focused onto the pinhole and is detected by a photodetector behind the screen. The detector is usually a photomultiplier tube (PMT). Light incident on the interferometer at an angle θ_r to the normal satisfies a modified resonance condition, given by $v_m = mc/(2d \cos \theta_r)$. However, because this light travels at an angle to the optic axis of the lens, it is focused to the side of the pinhole at point P_2 and is not detected. When the mirror separation is changed linearly with time and the output of the PMT is recorded on a strip chart recorder, the frequency content of the laser is displayed. From a measure of the frequency spread Δv in the laser light, the coherence length can be estimated. As we shall see in Chapter 4, other significant features of the laser output can also be observed.

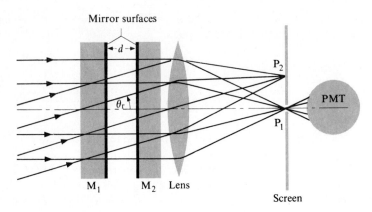

Fig. 2.19 Schematic of a Fabry-Perot interferometer used to examine the spectral content of a light source. The mirror supports and a mechanism for changing the mirror separation, d, are not shown. Light parallel to the optic axis of the lens and at the resonant frequency for the mirror spacing is transmitted by the Fabry-Perot cavity, focused onto a pinhole at point P_1, and detected by the photomultiplier (PMT). Not all Fabry-Perot interferometers have flat mirrors. Those with spherical mirrors have increased resolution for the same mirror separation, but the separation is fixed by the radii of curvature of the mirrors.

The Fabry-Perot resonator is very useful in measuring and displaying the properties of laser light. Its importance to the laser field goes much deeper than that, however, since the laser mirrors themselves form an optical resonator whose resonance conditions are identical in form to Eq. (2.31). Our discussion here will be useful later when we investigate the laser itself.

The Michelson Interferometer

A more direct measurement of the coherence length of light can be obtained using a Michelson interferometer, shown schematically in Fig. 2.20. For coherence length measurements, light from the source (assumed small) is first collimated by a lens and then divided by a partially silvered mirror, or beamsplitter, into two beams of equal irradiance. These two beams travel in different directions, one to mirror M_1, the other to mirror M_2, where they are reflected back toward the beamsplitter (BS). Since some of the light is reflected by the beamsplitter and some is transmitted, the beamsplitter serves to recombine the beams, forming once again a single beam. This beam is incident on the final output screen. In traveling from the beamsplitter to the mirrors and back again, the light in the two separate beams travels different pathlengths. The one-way pathlength difference from beamsplitter to mirrors is $L_1 - L_2$, where L_1 and L_2 are the distances from the center of the beamsplitter to mirrors M_1 and M_2, respectively. The total round-trip path difference is therefore $2(L_1 - L_2)$.

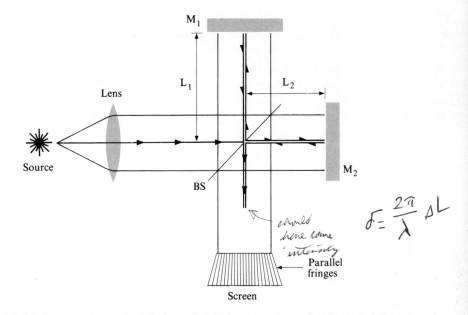

Fig. 2.20 Michelson interferometer. The beam is split into two parts, of pathlength difference $\Delta L = 2(L_1 - L_2)$, and recombined on the final output screen. To produce the parallel fringe pattern, one of the mirrors is tilted slightly.

To understand the operation of the Michelson interferometer in this application, assume that $L_1 = L_2$ initially (i.e., that the optical pathlengths are identical). The mirrors are misaligned slightly such that the beams, upon recombining, are not perfectly parallel to each other. Incident on the output screen, then, are two superposed plane-wave segments making a small angle to each other, just as in Young's experiment. Since the two waves originate at the same source and travel the same distance in reaching the screen, they exhibit complete coherence, and their superposition results in an interference pattern consisting of a set of parallel sinusoidal fringes. If the two waves have equal amplitude, the fringe visibility is unity.

Suppose now that we move one of the mirrors, introducing a nonzero pathlength difference $\Delta L = 2(L_1 - L_2)$. As this difference increases, the coherence between the interfering waves decreases, and the fringe visibility drops off. The interfering waves still originate from the same source, but they were emitted at different times. For a truly monochromatic source, this delay would have no effect on the fringe pattern (other than to displace it on the output screen), since the interfering waves would differ only in a constant relative phase angle. For real light sources, however, the change in pathlength means that the interfering waves no longer have the same amplitude and relative phase. The light from a real source, even a very narrow bandwidth source, must be described by electric field variations of the form $E(t) = A(t) \cos [\omega t + \varphi(t)]$, where both $A(t)$ and $\varphi(t)$ fluctuate *randomly* in time. The rate at which $A(t)$ and $\varphi(t)$ change depends on the bandwidth of the source. Even for a laser, they can change substantially over an interval of 10^{-8} seconds or less, and for most sources more rapidly than that. It is primarily the time-varying phase angle, $\varphi(t)$, that determines the temporal coherence characteristics of the wave, and, therefore, we concentrate on its influence in determining fringe visibility, ignoring the variations in amplitude.

Since the pattern on the output screen results from two interfering plane wave segments, we refer back to our earlier analysis of Young's experiment. The only change necessary is to replace φ_1 and φ_2 in Eqs. (2.17) and (2.18) with $\varphi(t)$ and $\varphi(t + \Delta t)$, where Δt is the differential time delay introduced by the mirror separation, i.e., $\Delta t = \Delta L/c$. The resultant expressions for the interfering waves are then

$$E_1(x,y,t) = A \cos [\omega t + \varphi(t)] \tag{2.34}$$

and

$$E_2(x,y,t) = A \cos [\omega t - kx \sin \theta + \varphi(t + \Delta t)] \tag{2.35}$$

where we have again assumed that the screen is in the $z = 0$ plane. One wave is simply a tilted and delayed version of the other. The resulting interference pattern, now varying in time, is given by

$$I(x,t) = CA^2[1 + \cos (kx \sin \theta + \Delta\varphi(t))] \tag{2.36}$$

where $\Delta\varphi(t) = \varphi(t) - \varphi(t + \Delta t)$. In obtaining this expression, we have simply substituted for $\Delta\varphi$ in Eq. (2.24). Implicit in such substitution is the assumption that the interval of the time average in Eq. (2.20), T, although large compared to $1/\nu$ (or $1/\omega$), is small compared to the reciprocal of the frequency bandwidth of the radiation, i.e., that $1/\nu \ll T \ll 1/\Delta\nu$.

The instantaneous relative phase of the two waves at the screen, $\Delta\varphi(t)$, determines the instantaneous position, or spatial phase, of the fringe pattern on the screen. For zero pathlength difference, $\Delta\varphi(t) = 0$, and the fringes are stationary. For nonzero pathlength difference, however, $\Delta\varphi(t)$ changes in time, and the fringe pattern moves back and forth in a random manner. The movement is much too rapid to be seen by the human eye, the rate of movement being determined by the bandwidth of the light (10^8 to 10^{14} Hz, depending on the source). What we see is the time average of the fringe pattern. So long as the pathlength difference is small, $\varphi(t)$ and $\varphi(t + \Delta t)$ are well correlated and $\Delta\varphi(t)$ remains small compared to 2π radians. The fringes move back and forth randomly, but the movements are small compared to the separation of the fringes. The time-average fringe visibility is still high, the coherence near unity. However, as the pathlength difference approaches the coherence length for the source, i.e., as ΔL approaches

l_c, the excursions become so great that the time-average fringe pattern is "washed out" and fringe visibility goes to zero.

This technique for measuring coherence length is employed with nonlaser sources and is even easier with laser sources. In the latter case, the coherence length is typically at least several centimeters, and careful mirror positioning is not required for the fringes to appear. The Michelson interferometer is a versatile tool, however, and it should not be construed that its applicability is limited to the measurement of coherence lengths. Particularly since the invention of the laser, its importance in the measurement of small distances or changes in distances, both in the research laboratory and in the industrial machine shop, has been very great.

2.7 POLARIZATION

Until now we have made no attempt to associate a direction with the electric field of a light wave. At any given point in space and instant in time, however, the electric field points in a particular direction, and is therefore appropriately described by a vector, E. This vector is perpendicular both to the propagation vector, k, which describes the direction of travel of the wave, and to the instantaneous direction of the magnetic field of the wave, H. As before, we concentrate on the electric field in our discussion and specify the direction of the electric field vector as the *direction of polarization* of the light. Light from incandescent bulbs, arc lamps, the sun, and many other sources is generally referred to as "unpolarized," or *randomly polarized*, meaning that the electric field vector associated with waves from these sources has no preferred direction of orientation; rather, it changes orientation quite rapidly in time, first pointing one way, then another. Certain other sources, however, produce light waves characterized by highly oriented electric fields. Such sources are generally referred to as "polarized" sources. Because many lasers produce light with highly oriented electric fields, we now add the concept of polarization to our list of properties of laser light. In this section, we will describe why laser beams exhibit this polarization and how light polarization can be used in the manipulation of laser beams.

Forms of Polarization

The simplest type of polarization is one for which the electric field orientation is constant with time, only the magnitude and sign of the field changing. This is known as *linear polarization*. If the electric field always points along the y-axis, the wave is said to be linearly polarized in the y-direction. The wave could just as well have been linearly polarized along any other direction in the plane normal to k. In Fig. 2.21, we illustrate the oscillations of the electric field for a light wave whose polarization direction is at some angle θ to the y-axis.

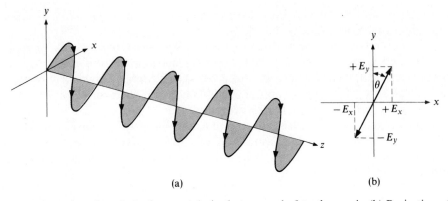

Fig. 2.21 (a) Linearly polarized wave polarized at an angle θ to the y-axis. (b) Projection of the electric field on a plane perpendicular to the z-axis.

As with any vector, the polarization vector can be split into its components, E_x and E_y, along orthogonal axes. We can describe linearly polarized light in two ways: by the maximum amplitude of the field and the angle the polarization vector makes with the y-axis, or by the x- and y-components of the maximum amplitude. Both ways are shown in Fig. 2.21(b). These descriptions are equivalent, their relation being the same as that between rectangular and plane polar-coordinate descriptions of the same vector. If the cartesian-coordinate description is used, the two components of the field must have the same phase, so that the maxima and minima for each component occur at the same time.

It is possible to have a light wave for which the x- and y-components of the electric field vector do not have the same phase. Under such circumstances, the amplitude maxima of the components do not occur at the same time. A direct consequence is that the wave is no longer linearly polarized, but exhibits a different form of polarization. Of particular interest is the polarization that results when the two components are 90° or 270° out of phase.* When the phase difference between components is 90° or 270°, the projection of the electric field vector on the x–y plane traces out an ellipse in time. Light having such an electric field vector dependence is termed *elliptically polarized*. Such a wave is illustrated in Fig. 2.22. Note that the electric field vector never passes through zero; there is always a nonzero electric field in an elliptically polarized wave. As the wave advances along the positive z-direction, the electric field vector rotates around the ellipse. The question is, how does it rotate, clockwise or counterclockwise?

---

*When the two components are 180° out of phase, the wave is again linearly polarized. Check this out for yourself.

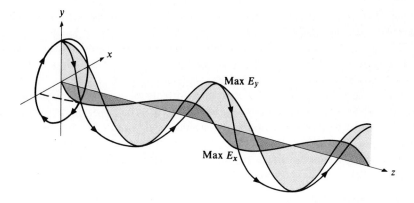

Fig. 2.22 Elliptically polarized light. The x-axis and y-axis components have been resolved. Note that the maximum of one components is located $\lambda/2$ behind the maximum of the other (90° out of phase). The path of the electric field vector as a function of time is projected on the x–y plane.

To describe the waves in a unique manner, two rules are used:

1. The electric field vector projection is always viewed with light propagating toward the observer (i.e., we look in the *minus* **k** direction).

2. The phase difference φ is taken as φ_y (phase of the y-component) minus φ_x (phase of the x-component).

If $\varphi = \varphi_y - \varphi_x = 90°$, then E_y leads E_x, and the vector in the x–y plane rotates *clockwise with time*. If $\varphi = 270°$ (or $-90°$), E_x leads E_y, and the vector rotates *counterclockwise with time*.

A special case occurs when the x- and y-components of elliptically polarized light are equal (i.e., $E_x = E_y$). In this case, the projection of the electric field vector on the x–y plane executes a circular path. This special case is called *circular polarization*. The sense of rotation with time in the x–y plane is exactly the same as before, since circular polarization is obviously a special case of elliptical polarization. For $\varphi = \varphi_y - \varphi_x = 90°$, the electric field vectors at one instant in time form a right-handed screw whose rotation with respect to the positive z-direction is counterclockwise.* Because the spatial form of a circularly polarized wave with $\varphi = 90°$ is a right-handed screw, the light is termed *right circularly polarized* (RCP).

---

* If at a particular point in space the wave rotates clockwise as time passes, the spatial dependence at a particular instant in time must be counterclockwise for a circularly polarized wave advancing in the positive z-direction, because the phase of such a wave is $\omega t - kz$.

Similarly for $\varphi = 270°$, the wave forms a left-handed screw and is called *left circularly polarized* (LCP).*

Elliptical polarization is more general than we have indicated here. In our description above, the major and minor axes of the ellipse were aligned along the *x*- and *y*-axes. In general, however, the orientation can be arbitrary. The decomposition of the rotating polarization vector into its *x*- and *y*-components is still possible, but the two components, in addition to having arbitrary magnitudes, now have arbitrary phase constants. It should be pointed out that although the figures suggest monochromatic light, monochromaticity is not required for light to be polarized. White light can be as easily polarized as light from a filtered gas discharge tube . . . or light from a laser.

Polarization Modifiers

There are several different methods for changing randomly polarized light to linearly polarized light. This is usually achieved by one of three general methods: (1) absorption, (2) reflection, or (3) transmission through birefringent crystals.

Polarization by absorption is most familiar to us, since this is the method of polarization used in polarizing sunglasses. In certain elongated molecules, light polarized parallel to the molecule length is absorbed, while most of the light polarized normal to this direction is transmitted. Thus if we send randomly polarized light, which has equal amounts of each component, through a material containing such molecules, the component parallel to the long molecular axis is absorbed, along with about one-third of the light with its polarization normal to that axis. The resulting transmitted light is linearly polarized, about one-third as intense as the incident light. The direction of the polarization of the transmitted light is called the *polarizer axis*. This direction is perpendicular to the long axis of the molecules.

If the light incident on a polarizer is already linearly polarized, then the amount transmitted by the polarizer depends on the angle θ between the electric field vector of the light wave and the polarizer axis. Let A be the amplitude of the light transmitted when the electric field vector and the polarizer axis are parallel. For nonparallel cases, only that component parallel to the polarizer axis is transmitted, with an amplitude $A \cos \theta$. The irradiance, being proportional to the time-averaged square of the electric field, then has a cosine squared dependence (the *law of Malus*):

$$I_{trans} = I_0 \cos^2 \theta \tag{2.37}$$

where I_0 is the maximum irradiance, transmitted when $\theta = 0$.

_____

* In case you have forgotten, all screws and bolts that we commonly use have a right-handed thread (except, of course, those machines you try to repair yourself and the lugs on the wheels on one side of your car. Which side we never remember.).

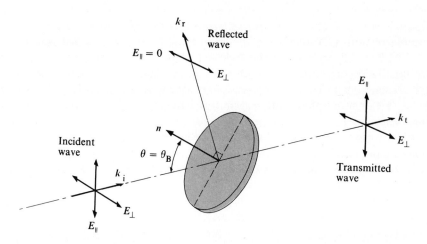

Fig. 2.23 Polarization by reflection. The vector n is the normal to the window surface; k^i is the incident light propagation direction. The reflected $E_{\parallel}$ component is zero when the window is oriented at the Brewster angle, $\theta_B = \tan^{-1} n$, where n is the refractive index of the window.

Polarization by reflection is the result of interaction of a light wave with a material surface. Light with its polarization vector parallel to the reflecting surface is preferentially reflected, as illustrated in Fig. 2.23. In this figure, we analyze randomly polarized light by considering it as two perpendicular vectors of equal length normal to the incident propagation vector $\mathbf{k}_i$. We also show the transmitted $(\mathbf{k}_t)$ and reflected waves $(\mathbf{k}_r)$ at some angle θ to the surface normal, $\mathbf{n}$. The plane containing both $\mathbf{k}_i$ and $\mathbf{n}$ is the *plane of incidence*, and the electric fields are labeled according to their polarization with respect to the plane of incidence. At normal incidence ($\theta = 0°$), about 4 percent of the light is reflected in all polarizations at one surface. In passing through a glass plate (two surfaces), about 8 percent of the light is reflected. For some general angle ($\theta \neq 0°$), most of the light polarized parallel to the plane of incidence is transmitted, while little of it is reflected. At an angle θ_B, known as the *Brewster angle*, the refractive indices of the two media that form the surface are such that all light that is polarized parallel to the plane of incidence is transmitted. It can be shown that the Brewster angle is related to the refractive index of the medium into which the light is transmitted, n_2, and to the refractive index of the medium of the incident light, n_1, by

$$\theta_B = \tan^{-1} \left(\frac{n_2}{n_1} \right). \tag{2.38}$$

If the medium of the incident light is air, then $n_1 \approx 1$, and $\theta_B \approx \tan^{-1} n_2$. By simply tilting a piece of glass at a particular angle to the direction of incident radiation, we can achieve 100-percent transmission (neglecting absorption) for one polarization. One method of polarizing light is to pass it through a series of glass plates all oriented at the Brewster angle. At each surface, some of the light polarized perpendicular to the plane of incidence is reflected, while all the light polarized parallel to the plane of incidence is transmitted. After passing through six or more plates, the transmitted light is highly polarized in the plane of incidence.

Polarization with birefringent crystals relies on the fact that light of different polarizations travels at different velocities in these crystals. Within the crystal, there are two directions perpendicular to each other known as the "fast" and "slow" axes of the crystal. Light polarized parallel to the fast axis has a higher velocity within the crystal than light polarized parallel to the slow axis. Equivalently, the refractive index for one polarization, n_{fast}, is not equal to that for the other polarization, n_{slow}. (Since v_{fast} is greater than v_{slow}, n_{fast} must be less than n_{slow}.) A beam of light incident on such a crystal is sorted into two components whose polarizations are parallel to the two axes. If the crystal is cut so that the second surface is at an oblique angle to the beams in the crystal (as for the left prism in Fig. 2.24), the two beams have different angles of refraction when they leave the first prism. However, if the angle between the beams and the normal to the second surface is large enough, the light of one polarization is totally internally reflected, while the other polarization is transmitted, thus separating the beams. A second crystal cut at the same angle redirects the transmitted beam parallel to its original direction. If the space between the prisms is filled with cement or oil, the refractive index of that material must be taken in account when calculating the prism angle.

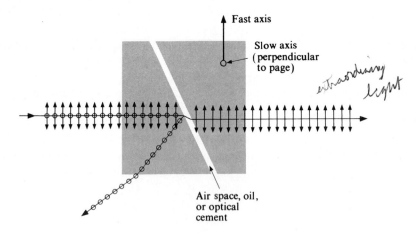

Fig. 2.24 Polarization by a birefringent crystal cut in the form of two prisms.

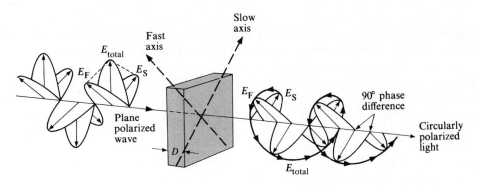

Fig. 2.25 Quarter-waveplate as a circular polarizer. Linearly polarized light with vertical polarization enters the crystal from the left. The velocities of the two components, E_F and E_S, are not equal inside the crystal. Emerging from the crystal (right), the E_S component lags the E_F component by 90°, and circular polarization results.

Another polarization modifier that uses a birefringent crystal is the *waveplate*. Suppose a linearly polarized beam, its plane of polarization at 45° to the fast and slow axes, enters a birefringent crystal of thickness D, as shown in Fig. 2.25. It takes the fast wave D/v_{fast} seconds to traverse the crystal. The slow wave emerges after D/v_{slow} seconds. The time lag between these two polarizations, which were originally in phase, is

$$\Delta t = \frac{D}{v_{slow}} - \frac{D}{v_{fast}} = \left(\frac{n_{slow}}{c} - \frac{n_{fast}}{c}\right) D \tag{2.39}$$

and the corresponding phase difference is

$$\Delta \varphi = 2\pi v \, \Delta t = \frac{2\pi v}{c}(n_{slow} - n_{fast}) D = \frac{2\pi}{\lambda_0}(n_{slow} - n_{fast}) D \tag{2.40}$$

where λ_0 is the wavelength in a vacuum. If the crystal is cut to a thickness D such that

$$(n_{slow} - n_{fast}) D = \frac{\lambda_0}{4}$$

the crystal is called a *quarter waveplate*. The phase difference is then

$$\Delta \varphi = \frac{2\pi}{\lambda_0} \cdot \frac{\lambda_0}{4} = \frac{\pi}{2}$$

and the two components, originally in phase, are now $\pi/2$ (90°) or one-quarter wave out of phase. Since the beams entered with equal amplitudes, the resultant output polarization is circularly polarized light. Changing the polarization vector from 45° on one side of the crystal axis to 45° on the other side of the axis results in a change in the handedness of the polarization. If an angle other than 45° is used, elliptical polarization results.

If the crystal thickness is doubled to give a *half waveplate*, the emerging beams are 180° out of phase with each other. The resultant polarization is still linear, but the polarization vector is rotated through some angle. When the angle between the fast crystal axis and the input polarization direction is θ, the angle of rotation of the emerging polarized wave is 2θ. Using such devices as the half and quarter waveplates, the laser output can be modified for amplitude modulation in optical communication systems and for Q-switching and mode locking, topics we discuss later.

Polarization of Laser Light

The output of many lasers is linearly polarized, with the ratio of the light polarized in one direction exceeding that polarized in the orthogonal direction by 1000:1. In most cases, this high degree of linear polarization is the result of a Brewster surface within the laser. Such a surface is commonly used in the construction of a laser because light must be transmitted out of the lasing medium with as little loss as possible. Cutting the ends of a laser rod or mounting windows of a gas laser tube at the Brewster angle assures that light of one polarization direction is transmitted out of the lasing medium to the reflecting mirrors and back into the lasing medium with no loss. These angle-cut surfaces and windows are called *Brewster cuts* or *Brewster windows*. For light polarized perpendicular to the plane of incidence, there is a large loss at the Brewster surface due to the reflection out of the lasing medium. Usually this loss makes it impossible for light of this polarization to lase. However, the preferred polarization can lase, accounting for the high degree of polarization of the output.

Some lasers are built without Brewster surfaces or Brewster cuts, and might be expected to produce unpolarized output. However, there is always some degree of birefringence or polarization selectivity associated with optical components in a laser. As a consequence, the output of even an "unpolarized" laser is polarized, consisting of two or more collinear beams of linearly polarized light of slightly different wavelengths. The polarization directions of these beams are orthogonal to one another and do not change with time. The output power associated with each polarization will vary slowly with time, however, as the laser warms up. Although this kind of polarization is not evident by passage of the light through a simple polarizer, the output of all lasers differs radically from the randomly polarized output of other light sources.

2.8 A NOTE ON COHERENCE

The high degree of coherence exhibited by laser light is intimately related to the laser's characteristics of brightness, directionality, monochromaticity, and even, in a certain sense, of polarization. We have already seen that the bandwidth of a laser is inversely proportional to its coherence length. And were it not for the high transverse coherence of laser light, the output of a laser could not consist of the highly directional planar wave segments described earlier. Since high spectral brightness results from simultaneous high directionality and small bandwidth, the laser, possessing both high spatial and high temporal coherence, is indeed bright.

Coherence is a manifestation of the great regularity—the great predictability in time and in space—of the light waves produced within a laser. We shall discuss the origins of this regularity in Chapter 4. We note here, however, that the regularity relates not only to the amplitude and phase of the waves, on a point-by-point basis, but also to the polarization. At every point, coherent light has a fixed direction of polarization. Only to the extent that light is not perfectly coherent can that polarization change with time. This relationship between coherence and polarization is not sufficient to guarantee linearly polarized light output from all lasers. Nonetheless, as noted earlier, the conditions within a laser that are conducive to the generation of coherent light are often also conducive to the generation of light with a specific state of linear polarization.

In summary, we see that all the properties of laser light we consider so special are, in one way or another, closely coupled with the coherence of that light. The one aspect of laser light that most clearly differentiates the laser from all other sources of light is, quite simply, coherence.

PROBLEMS

2.1 Complete the table given below. Where only the region is given, pick a value for λ or v near the center of that region. Be sure to specify your units.

Region	Wavelength	Frequency	Frequency bandwidth, Δv (for $\Delta \lambda = 0.1$ nm)	$\dfrac{\Delta v}{v} \times 100\%$
	1 m			
		115 GHz		
				0.006%
Visible				
	176 nm			
		3×10^{18} Hz		

2.2 A ring laser consists of three or four lasers arranged in a closed figure, usually a triangle or square. What are the propagation vectors for light in an equilateral triangular helium-neon ring laser if it is made of tubes of equal length, with one side parallel to the x-axis of a coordinate system?

2.3 Given the expression for a traveling wave as $y = 5 \sin (6\pi t - 10\pi x)$, x in cm, t in seconds, what is

   a) the amplitude?
   b) the wavelength?
   c) the angular spatial frequency?
   d) the frequency?
   e) the angular temporal frequency?
   f) the speed of propagation?
   g) the direction of propagation?

2.4 If the criterion for a point source given in the box in Section 2.1 is modified to an equality by replacing the $\ll$ by a factor of 10 (i.e., $10H\Phi = \lambda$), what are the acceptable sizes of sources if (a) you are working in a laboratory (assume you have a 4-meter optical bench), (b) you are working in the atmosphere at a range of 7 km, (c) the source is on the moon (earth-moon distance: 250,000 miles)? Assume $\lambda = 500$ nm.

2.5 A laser has a full angle beam divergence of 0.2 milliradians.

   a) If the beam cross section is circular, what is the solid angle of the beam?
   b) What fraction of 4π steradians does this solid angle represent?
   c) If the power of the beam is 10 milliwatts, what is the irradiance at a point 25 meters distant from the laser?
   d) How powerful would a source have to be to provide the level of irradiance calculated in part (c) at 25 meters in all directions?

2.6 The moon is one-quarter million miles away. What is the surface area illuminated by a laser with a full angle beam divergence of (a) 1 mrad? (b) 1 μrad? (c) The divergence of a laser beam can be reduced by expanding and recollimating the beam at a larger diameter, the divergence being calculated using Eq. (2.15). What must the diameter of an expanded argon-ion laser beam be for a divergence of 1 μrad? Assume the laser operates at $\lambda = 514.5$ nm.

2.7 A laser beam ($\lambda = 500$ nm) uniformly illuminates a circular aperture inside a long tube. The aperture cannot be measured, but the circular diffraction pattern at the other end of the tube can be measured.

   a) If the aperture is 2 m from the edge of the tube and if the distance from the *center* of the diffraction pattern to the first dark ring is 6.1 mm, what is the angular diameter of the first dark ring?
   b) Given the answer to part (a), what is the diameter of the aperture?

2.8 If the angle between two interfering beams as shown in Fig. 2.12 is 0.8 milliradians and the wavelength of light is 633 nm, what is the distance between two successive bright fringes on the screen?

2.9 The optical system shown in Fig. 2.16 can be adjusted for various pathlength differences. If the distance between beamsplitters equals 5 cm and X_D is 20 cm,

a) what is the pathlength difference?
b) what is the time delay introduced into one wave by the longer path?
c) what is the phase difference introduced between the two waves by this geometry, if $\lambda = 5$ cm?
d) write the expression for the wave at the second beamsplitter (BS_2), if $y = A \cos(\omega t - kz)$ is the expression for the wave at the first beamsplitter (BS_1).
e) is the interference at beamsplitter BS_2 constructive or destructive $(\lambda = 2$ cm$)$?

2.10   What is the coherence length of a source $\lambda = 500$ nm with a 1-nm bandwidth? What must the bandwidth be to obtain a 1-m coherence length?

2.11   A Fabry-Perot interferometer has a plate spacing of 4.5 cm. Calculate the free spectral range. If the HeNe laser line at 632.8 nm has a linewidth of about 0.01 nm, can a scanning Fabry-Perot interferometer display the lineshape?

2.12   A Fabry-Perot interferometer is set at exactly 2.0000 mm mirror separation. What is the wavelength closest to 514.5 nm that will be transmitted? What is the integer m for this transmission? What is the wavelength λ_2 of the $(m + 1)$th transmission? What is the frequency separation between these two transmissions? By how much and in what direction must the spacing change for λ_2 to be the mth transmission? (Be accurate; you are dealing with small differences between large numbers.)

2.13   A beam of laser light is polarized in the vertical plane. A polarizer P_1 is placed in the beam with its axis horizontal. What percent of the light is transmitted if P_1 is a perfect polarizer? Now, a second perfect polarizer P_2 is inserted in the beam *between* P_1 and the laser. Its axis is at 45° to the vertical. What percent of light is transmitted now? Plot a graph of percent light transmitted versus the angle of P_2 between vertical (0°) and horizontal (90°).

2.14   What is the Brewster angle for a laser window made of (a) fused quartz $(n = 1.462)$? (b) glass $(n = 1.523)$? (c) plastic $(n = 1.421)$?

2.15   If polarizer sunglasses cut "glare," and glare is defined as light reflecting off a surface and preferentially polarized parallel to the surface, how are the polarized layers in the sunglasses oriented?

2.16   Given Eq. (2.38) for the Brewster angle, show that the reflected and refracted beams are at right angles to each other.

REFERENCES

2.1 A. P. French (1971), *Vibration and Waves*. New York: W. W. Norton. Everything you have always wanted to know about wave motion.

2.2 F. A. Jenkins and H. E. White (1976), *Fundamentals of Optics* (4th edition). New York: McGraw-Hill. For many years the only optics book in general use.

2.3 M. V. Klein (1970), *Optics*. New York: Wiley. Treatment of physical optics is above the level of Chapter 2. Useful because it discusses the Fabry-Perot interferometer and coherence.

2.4 M. R. Meyer-Arendt (1972), *Introduction to Classical and Modern Optics*. Englewood Cliffs: Prentice-Hall. A wide-ranging text with discussion of many modern optical techniques. Treatment at or above the level of Chapter 2.

2.5 J. Strong (1958), *Concepts of Classical Optics*. San Francisco: W. H. Freeman. A text that can be enjoyed through casual reading. The wide-ranging topics of the appendixes and Roger Hayward's drawings add to the enjoyment.

2.6 A. C. S. van Heel and C. H. F. Velzel (1968), *What is Light?* New York: McGraw-Hill. A delightful, nonmathematical discussion of optics. (Paperback)

2.7 R. H. Webb (1969), *Elementary Wave Optics*. New York: Academic Press. An excellent text in physical optics at approximately the level of Chapter 2.

3

Basic Laser Principles

In order for most lasers to operate, three basic conditions must be satisfied. First, there must be an *active medium*, that is, a collection of atoms, molecules, or ions that emit radiation in the optical part of the electromagnetic spectrum. (We frequently speak of the active medium as though it were a collection of atoms in a gas. Just as easily, however, the active medium of a laser might be atoms, molecules, or ions in liquid or solid form.) Second, a condition known as a *population inversion* must exist. This condition is highly abnormal in nature. It is created in a laser by an excitation process known as *pumping*. Finally, for true laser oscillation to take place, there must be some form of optical *feedback* present in the laser system. If this were not present, the laser might serve as an amplifier of narrowband light, but it could never produce the highly collimated, monochromatic beam that makes the laser so useful. These three basic conditions and their underlying principles are the central topic of this chapter. Later in Chapter 6, we discuss actual materials and devices used in satisfying these conditions in a number of different lasers.

3.1 THE ACTIVE MEDIUM: THE ATOMIC BASIS FOR LASER ACTION

As early as the 1750's, evidence was being accumulated that would eventually support the quantum theory of light and matter. Gas discharge tubes, the forerunners of today's neon signs, were found to emit light only at discrete wavelengths rather than in the continuous spectra predicted by classical theories. When light from these sources was passed through a slit and dispersed by a prism, the spectra of atomic elements excited by electrical discharge appeared as groups of lines. Similarly, it was found that many gases, when illuminated by a white light source, exhibited substantial absorption of light only at discrete wavelengths. Because of their special characteristics, emission and absorption spectra such as these came to be known as *line spectra*, and by 1823 it had been shown that each atomic element produced a characteristic and identifiable line spectrum. Later, molecules were seen to exhibit the same kind of behavior, emitting or absorbing radiation at discrete wavelengths in the infrared part of the spectrum.

Discrete Energy States and Transitions

The explanation of line spectra had to wait until 1913, when Niels Bohr developed a theory that allowed him to predict the wavelengths of the lines in the simplest of these spectra, that of hydrogen. Bohr assumed that atoms can exist for long periods of time in certain discrete energy states and postulated that the observed emission or absorption of light by those atoms occurs when an atom makes a transition from one discrete energy state to another. The frequency of the emitted or absorbed radiation, v, is related to the difference in energy of the two states by the expression

$$\Delta E = hv \tag{3.1}$$

where $h = 6.625 \times 10^{-34}$ joule-sec. The constant of proportionality, Planck's constant, had previously appeared in discussions of the quantum theory of blackbody radiation.

Bohr's picture of the atom—a central nucleus orbited by a charged electron—was based on classical concepts. Each energy state corresponded to the electron orbiting at a discrete radial distance from the nucleus. The development of modern quantum theory has changed this description. The classical concept of an electron orbiting a nucleus like a planet about its sun has been replaced by a probabilistic description, which states where the electron is *most likely* to be found but does not specify its exact position. This probabilistic description is contained in the atomic *wave function*, which is the solution to a differential equation (the Schroedinger equation) that describes the atomic system's dynamic characteristics and contains all the necessary information about the electron's environment: the charge of the nucleus, the electric field arising from the presence of other electrons, a description of any external electric or magnetic fields, and so forth. During the past half-century, quantum theory has been developed to a highly refined mathematical discipline. Nonetheless, the relatively simple concepts of stable energy states and radiative transitions between these states remain valid. An atom in some arbitrary energy state E_i can be induced by radiation at frequency v_{ij} to undergo a transition to a higher energy state E_j if $v_{ij} = (E_j - E_i)/h$. This is the essence of the absorption process. Emission is the reverse process: The atom, initially in state E_j, emits a quantum of radiation, or *photon*, of energy hv_{ij}, and in so doing makes a transition to state E_i.

Quantum Numbers

The wave-function solution to the Schroedinger equation for an electron in orbit around a central nucleus and the stable electronic energy state, or *quantum state*, that it describes have as parameters four numbers, called *quantum numbers*. Each of these numbers, which are designated by specific symbols, is associated with a particular physical property of the atomic system. The associated physical properties, the conventional notation, and the range of allowable values for these quantum numbers are shown in Table 3.1.

Table 3.1 Quantum numbers.

Name	Symbol	Physical property	Possible values
Principal	n	Size of electron orbit	$1, 2, 3, \ldots$
Orbital	l	Angular momentum*	$0, 1, 2, \ldots, n-1$
Magnetic	m	z-component of angular momentum†	$-l \leqslant m \leqslant l$
Spin	s	Electron spin	$+1/2, -1/2$

* For historical reasons, the angular momentum quantum number is usually represented by a letter according to the scheme

Value of l	0	1	2	3	4	5	6	$\ldots$
Spectroscopic symbol	s	p	d	f	g	h	i	$\ldots$

A symbolic notation for atomic states results. For example, $3p$ signifies the state $n = 3$, $l = 1$; $2s$ signifies $n = 2$, $l = 0$. The symbol, s, indicating $l = 0$, should not be confused with that for the electron spin. Its usage is usually clear from context.

† The quantum number relating to the z-component of angular momentum, m, affects the energy of the electron only when some preferred direction, the z-axis, has been established by the application of an external magnetic field.

Each quantum state for the one-electron atom is characterized by a unique set of quantum numbers (n, l, m, s). The quantum number n is called the *principal quantum number* because the differences in energy are greatest for electrons in quantum states with different n values. It might be referred to as a radial quantum number, since it is directly related to the average distance of the electron from the nucleus. The quantum numbers l and m determine (in a probabilistic sense) the extent to which the electronic orbit deviates from a perfect sphere. Different values of l, associated with differing amounts of orbital angular momentum, correspond to much smaller differences in energy. The electron spin, designated by s, has no classical counterpart but can contribute to the energy of the atom when the electron is in a magnetic field. Atoms with the same values of n and l but different values of m and s often have the same total energy. Such quantum states are described as *degenerate energy states*. When atoms in these states are placed in a magnetic field, the energy of each degenerate state increases or decreases by an amount that depends on the values of m and s. The degeneracy is said to be removed by the presence of the field.

In multi-electron atoms, each electron is associated with a set of quantum numbers. A very important characteristic of atoms is expressed by the *Pauli exclusion principle*, which states that no two electrons in an atom can have the same set of quantum numbers. If one electron in orbit about a nucleus is characterized by the quantum numbers $n = 1$, $l = 0$, $m = 0$, $s = -1/2$, the second electron must have at least one quantum number different, for example, $s = +1/2$.

Selection Rules and Transition Lifetimes

The change of atomic energy state with the simultaneous absorption or emission of light is referred to as an *atomic transition*. Like electronic orbits, transitions between energy states are also described in probabilistic terms. Certain transitions, with their accompanying absorption or release of energy, are highly probable; these are referred to as *allowed transitions*. Certain other transitions, referred to as *forbidden transitions*, are highly improbable, occuring perhaps a thousand times less frequently than allowed transitions (and perhaps never!). The rules that govern transitions between quantum states and that consequently determine whether a transition is allowed are called *selection rules*.

One very simple example of a selection rule states that the total spin of an atom, obtained by summing the spins of its individual electrons, cannot change during a transition. Consider the helium atom as an example. In its lowest energy state, commonly called the *ground state*, the two electrons in a helium atom have spins of opposite sign.* The *total spin* of an atom, the sum of the individual electron spins, designated by S, is then $S = 1/2 + (-1/2) = 0$. In an excited (higher) energy state, the two electrons typically have different l-values, and the electron spins can be either alike or opposite in sign without violating the exclusion principle. The selection rule governing total spin then implies that transitions from the excited state to the ground state are allowed only if the spins of the electrons in the excited state are of opposite sign, or "paired." A two-electron atom in a state with paired electrons ($S = 1/2 - 1/2 = 0$) is said to be in a *singlet state*. If the spins are alike ($S = 1/2 + 1/2 = 1$), the atom is said to be in a *triplet state*. Thus, the selection rule tells us that transitions between singlet states or between triplet states are allowed. Transitions from triplet to singlet, or from singlet to triplet, however, are forbidden.

A measure of the probability of a transition is the *transition lifetime*, τ. This lifetime is considered to be either the period of time over which the transition takes place (i.e., the period of time over which the quantum of energy, $h\nu$, is absorbed or released) or, equivalently, the period of time during which the solution to the Schroedinger equation shows the atom is most likely to be found in the initial energy state of the transition. Typically, allowed transitions take place over periods of a microsecond or less. Because the atomic system may remain in the upper state of a forbidden transition for a thousand to a million times longer

---

* This satisfies the exclusion principle, since all other quantum numbers are the same (i.e., $n = 1$, $l = 0$, $m = 0$ for both electrons).

than in the upper state of an allowed transition, the upper states of forbidden transitions are often referred to as *metastable states.*

The various selection rules are generally obeyed by atoms with relatively small atomic numbers and by low-mass molecules. As atomic and molecular structures become more complicated, however, quantum-number assignments become more difficult to determine, and spectacular "violations" of the selection rules occur. Nevertheless, metastable states are found in atoms and molecules of large and small electron numbers alike—a fact of great importance to the development of the laser.

Absorption and Emission

With the preceding description of atomic quantum states and transitions as a basis, we can now consider in greater detail the absorption and emission processes that occur in collections of atoms or molecules. Consider an absorption spectrometer, which is equipped with a glass cell filled with an atomic gas. For simplicity, we assume that the atoms in the gas are characterized by two energy states: the ground state with an energy E_0, and a single excited state with an energy E_1 ($E_1 > E_0$). A beam of white light is passed through the cell and dispersed by the prism. A spectrum taken without the gas present in the cell is that of the white light source, a smoothly varying distribution of the colors of the spectrum. With the gas present in the cell the spectrum shows a dark line in the white light spectrum at a wavelength given by

$$\lambda = \lambda_{01} = \lambda_{10} = \frac{c}{v_{10}} \tag{3.2}$$

where

$$v_{10} = v_{01} = \frac{E_1 - E_0}{h} = \frac{\Delta E_{10}}{h}, \qquad E_1 > E_0$$

in accord with Eq. (3.1). The atoms in the gas have absorbed quanta of radiation energy at frequency v_{10}. Specifically, atoms have been stimulated by the incident radiation at frequency v_{10} to undergo transitions from the ground state to the excited state. Energy is removed from the light beam with a corresponding reduction in light level on the output screen for the transition wavelength, λ_{10}. Because incident light energy is necessary for this absorption process to occur, the process is often referred to as *stimulated absorption.*

But what of the atoms in energy state E_1? Does the absorption process continue until there are no atoms left in the ground state? This, in fact, does not happen, because emission processes quickly act to refill the ground state (natural systems tend to seek out the lowest energy configuration). Soon after being excited to the higher energy state, an atom emits a quantum of energy at frequency v_{10} and returns to the ground state. This emission process can assume two forms.

In contrast to an absorption transition, which must be induced by incident light energy, the emission of photon energy by an excited atom can occur spontaneously or it can be stimulated to occur by the presence of radiation at the transition frequency. Accordingly, the emission process is in reality two distinct processes: *spontaneous emission*, which occurs without any external stimulation, or *stimulated emission*, which occurs when the atom in its excited state interacts with an incident quantum of light energy at the transition frequency. These two independent processes, along with stimulated absorption, are illustrated schematically in Fig. 3.1.

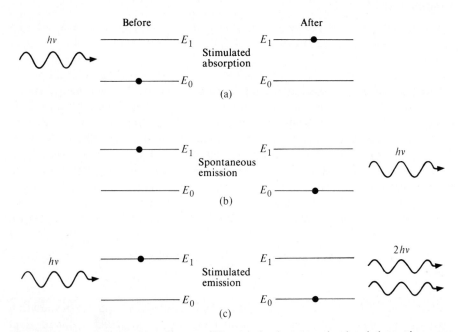

Fig. 3.1 Energy-state-transition diagram differentiating between stimulated absorption, spontaneous emission, and stimulated emission. A black dot indicates the state of the atom before and after the transition takes place. In the stimulated emission process, energy is added to the stimulating wave during the transition; in the absorption process, energy is extracted from the wave.

The stimulated emission process is an essential part of the operation of a laser and, in fact, lends the third and fourth letters to the acronym LASER (Light Amplification by Stimulated Emission of Radiation). Unlike atoms emitting radiation by spontaneous emission transitions, which occur randomly in time, atoms undergoing stimulated emission radiate their quanta of energy *in phase* with the stimulating radiation. This means that if an atom is stimulated to emit light energy by a propagating wave, the additional quantum of energy liberated in the process *adds to that wave on a constructive basis*, increasing its amplitude. It is as though one of the hypothetical secondary sources of Huygens' principle were augmented by a true source of radiation emitting in phase with the original wave.

Returning to our experiment with the gas cell, we have noted that a portion of the incident radiation at wavelength λ_{10} is absorbed by the atoms of the gas. Some small fraction of this energy is reradiated through stimulated emission, reinforcing the light beam. The major portion of energy in the excited atoms is released through spontaneous emission, however, radiating isotropically in all directions from the cell. The reduction in beam irradiance at ν_{10} is thus predominantly due to the absorption and subsequent spontaneous emission of wave energy.

Einstein Relations

The three transition processes—stimulated absorption, spontaneous emission, and stimulated emission—all occur in a collection of atoms. In 1917, Albert Einstein showed that the rates of these processes are related mathematically. The average fractional rate at which atoms in energy state E_1 spontaneously drop to the lower energy state E_0 is designated by a constant, A_{10}, the *Einstein coefficient of spontaneous emission*. The reciprocal of A_{10} is the time for the $(1 \rightarrow 0)$ transition to occur and therefore is the transition lifetime of the state we referred to earlier ($\tau_{10} = 1/A_{10}$). If there are N_1 atoms in the collection with energy E_1, the number of atoms per second that undergo spontaneous emission is given by the expression

$$\text{Spontaneous emission transition rate} = N_1 A_{10}. \qquad (3.3)$$

The value of A_{10}, and hence the total rate, depends upon the kind of atom and upon the particular transition involved.

For stimulated absorption and stimulated emission to take place, there must be radiation energy present. Einstein postulated that the number of atoms undergoing absorption transitions each second is proportional to the number of atoms in the lower energy state, N_0, and to the spectral density of radiation energy, ρ_ν (joule-sec/m^3), at the transition frequency, ν_{10}.

$$\text{Stimulated absorption transition rate} = N_0 \rho_\nu B_{01} \qquad (3.4)$$

where the constant of proportionality, B_{01}, is the *Einstein coefficient of stimulated absorption*. Similarly, the rate at which the stimulated emission process takes place, measured in transitions per second, is given by the number of atoms in the excited state times the spectral radiation density times the *Einstein coefficient of stimulated emission*, B_{10}:

$$\text{Stimulated emission transition rate} = N_1 \rho_\nu B_{10}. \qquad (3.5)$$

The effective spectral radiation energy density associated with the light beam is related to the spectral irradiance of the beam by $\rho_v = I_v/c$. In rewriting Eqs. (3.4) and (3.5) in terms of the spectral irradiance of the light beam, we thus obtain

$$\text{Stimulated absorption transition rate} = N_0 \frac{I_v}{c} B_{01} \tag{3.6}$$

and

$$\text{Stimulated emission transition rate} = N_1 \frac{I_v}{c} B_{10}. \tag{3.7}$$

If conditions of thermal equilibrium are assumed, the coefficients of stimulated absorption and stimulated emission can be shown to be equal,

$$B_{01} = B_{10} \tag{3.8}$$

and the coefficient of spontaneous emission can be shown to be related to the coefficient of stimulated emission by the formula

$$A_{10} = \frac{8\pi h v_{10}^3}{c^3} B_{10}. \tag{3.9}$$

Equations (3.8) and (3.9) are known as the *Einstein relations*. Although obtained under conditions of thermal equilibrium, the equations can be assumed to hold for steady-state (i.e., non-time-varying) conditions, even when true thermal, or blackbody, equilibrium is not established. The second Einstein relation states that the ratio of the probability of spontaneous emission to the probability of stimulated emission for a given pair of energy levels is proportional to the cube of the frequency of the transition radiation. This cubic dependency on v accounts for the principal difficulty in achieving laser action at x-ray wavelengths, where v is of the order of 10^{16} Hz. At this frequency, spontaneous emission occurs so rapidly for a high-energy transition that light amplification by the stimulated emission process is not easily achieved. Fortunately, this difficulty is not great for most lasers at visible wavelengths.

Our discussions have been in terms of transitions between the ground state and a single excited atomic state. Real atoms are characterized by a multiplicity of different energy states, however, and absorption and emission transitions are possible (in principle, at least) between any of these states. For such real atomic systems, the Einstein relations and the energy-frequency relation of Eq. (3.1) can be extended for an arbitrary pair of atomic energy states, E_i and E_j $(E_j > E_i)$, simply by substituting E_j for E_1 and E_i for E_0. Although the interrelation of rates for all possible transitions between atomic states will be quite complicated, they still reduce to the relatively simple Einstein relations on a pairwise basis.

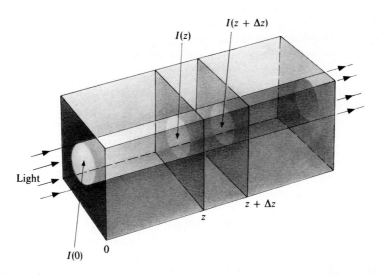

Fig. 3.2 Collimated beam of light traversing an absorbing gas. The change in the irradiance across a small slab of the gas is proportional to the irradiance at the slab and to the thickness of the slab, Δz.

Absorption and Small-Signal Gain Coefficients

Consider a collimated beam of light traveling in the z-direction and passing through an atomic gas, as shown in Fig. 3.2. For simplicity, we assume that there is only a single radiative transition, which occurs between two energy states E_i and E_j $(E_j > E_i)$ and that the incident light is monochromatic at the transition frequency, $\nu_{ij} = (E_j - E_i)/h$. The light in the beam is characterized by its irradiance, I,* as a function of position along the z-axis. We consider the change in $I(z)$ in traversing a small distance z between planes at z and $z + \Delta z$ in the gas,

$$\Delta I(z) = I(z + \Delta z) - I(z). \tag{3.10}$$

For a homogeneous gas, $\Delta I(z)$ is proportional both to Δz (for Δz sufficiently small) and $I(z)$. The constant of proportionality is, by definition, the *absorption coefficient*, α:

$$\Delta I(z) = -\alpha I(z) \, \Delta z. \tag{3.11}$$

The minus sign indicates the reduction in beam irradiance brought about by absorption, since α is a positive quantity. If we rewrite Eq. (3.11) in differential

---

* Because we assume a perfectly monochromatic wave, I represents the total irradiance, as opposed to the spectral irradiance used in Eqs. (3.4) to (3.7).

equation form, we obtain

$$\frac{dI(z)}{dz} = -\alpha I(z). \tag{3.12}$$

We can easily solve this equation for $I(z)$ as a function of the distance the light beam travels through the gas, with the result

$$I(z) = I(0)e^{-\alpha z} \tag{3.13}$$

which is known as *Beer's law*. $I(0)$ is the irradiance of the beam at the input to the gas cell.

ΔI is usually negative, corresponding to the normal attenuation of the incident beam. Under certain circumstances, however, ΔI can be positive; an incident beam of light at the correct frequency can be amplified, growing in magnitude as it travels through the slab. This amplification can be thought of as a negative absorption process. Prior to the invention of the laser, such amplification had never been observed. It is not difficult to suggest, however, how it might occur. Assume, for example, that all the atoms in the gas volume element are in the upper energy state, E_j. There can then be no absorption. There can, however, be stimulated emission: An incident wave of frequency ν_{ij} can stimulate atoms in the collection to radiate their stored energy; the energy is radiated at the same frequency, ν_{ij}, and in phase with the stimulating wave, thereby amplifying it.

Not all atoms need be in the upper energy state for amplification to occur; it is only necessary that the number of atoms in the upper energy state of the transition exceed the number in the lower state. In what follows, we use the Einstein relations to show that the absorption coefficient, α, is proportional to the difference in the populations of the two energy states and hence becomes negative if the population difference becomes negative. By equating negative absorption with amplification, we then show that having a majority of atoms in the upper of the two transition states is a necessary condition for light amplification by stimulated emission.

We refer once again to Fig. 3.2 and use Eqs. (3.6) and (3.7) to write an expression for the net number, n_{ij}, of energy quanta (photons) lost per second to the collimated beam as it travels through a volume element of gas of thickness Δz:

$$-\frac{dn_{ij}}{dt} = N_i \frac{I(z)}{c} B_{ij} - N_j \frac{I(z)}{c} B_{ji}$$

$$= B_{ij}(N_i - N_j)\frac{I(z)}{c} \tag{3.14}$$

where N_i and N_j are the number of atoms in the lower and upper energy states, respectively, in the volume element, and B_{ij} and B_{ji} are the stimulated absorption and stimulated emission coefficients, respectively, for the transition between these two well-defined energy levels.* B_{ji} has been eliminated in the second expression by applying Eq. (3.8). In writing this expression, we have ignored any scattering losses. Spontaneous emission has also been ignored, because such light, emitted in all directions, contributes little to the collimated beam.

To provide the link between this expression, which contains the difference in populations, and the absorption coefficient, we also express the rate of beam photon loss in terms of the values of the beam irradiance at z and at $z + \Delta z$:

$$-\frac{dn_{ij}}{dt} = [I(z) - I(z + \Delta z)]\frac{A}{hv_{ij}} \tag{3.15}$$

where $I(z)$ equals the beam irradiance at plane z, $I(z + \Delta z)$ equals the beam irradiance after passing through the gas volume element, and A equals the cross-sectional area of the beam (assumed to be uniform). We can equate expressions (3.14) and (3.15) for $-dn_{ij}/dt$ to obtain

$$[I(z) - I(z + \Delta z)]\frac{A}{hv_{ij}} = B_{ij}(N_i - N_j)\frac{I(z)}{c}. \tag{3.16}$$

If Δz is sufficiently small, we can rewrite Eq. (3.16) in the form

$$-\frac{dI(z)}{dz}\Delta z \frac{A}{hv_{ij}} = B_{ij}(N_i - N_j)\frac{I(z)}{c}. \tag{3.17}$$

Substituting Eq. (3.12) into this expression, dividing both sides by $I(z)$, and solving for α, we obtain

$$\alpha = B_{ij}\frac{(N_i - N_j)}{A\,\Delta z}\frac{hv_{ij}}{c} = B_{ij}(\mathcal{N}_i - \mathcal{N}_j)\frac{hv_{ij}}{c} \tag{3.18}$$

where we have let $\mathcal{N}_i = N_i/A\,\Delta z$ and $\mathcal{N}_j = N_j/A\,\Delta z$ represent the number of atoms per unit volume, or atomic *population densities*, of the two energy states.

Equation (3.18) provides the link between the absorption coefficient, α, and the population difference for the two transition states. Normally, $\mathcal{N}_i$ is greater than $\mathcal{N}_j$ in Eq. (3.18); α is a positive quantity and absorption occurs. Suppose, however, that we create a condition where $\mathcal{N}_j$ is greater than $\mathcal{N}_i$. Then α is negative, the quantity $(-\alpha z)$ in the exponent of Eq. (3.13) is positive, and the irradiance of the beam grows with distance according to the relation

$$I(z) = I(0)e^{\beta z} \tag{3.19}$$

---

* These are not precisely the Einstein coefficients defined previously, since we have suppressed any frequency dependence with our assumption of monochromatic radiation. In Section 4.1, we will include the frequency dependence of the processes discussed here.

where β, the so-called *small-signal gain coefficient*, is given by the relationship

$$\beta = -\alpha, \quad \alpha < 0$$

or

$$\beta = B_{ij}(\mathcal{N}_j - \mathcal{N}_i)\frac{h\nu_{ij}}{c}. \tag{3.20}$$

The light beam is amplified rather than attenuated. The process causing this growth in beam irradiance, light amplification by stimulated emission, is the underlying basis for laser operation.

As noted, the number of atoms in the lower energy state normally exceeds the number in the upper state, and α is positive (or β is negative). It is possible by a process known as *pumping* to reverse the situation and to create a condition where the population of the upper energy state exceeds that of the lower. The attainment of this condition, known as *population inversion*, constitutes the second principal requirement of a laser.

3.2 LASER PUMPING: CREATING A POPULATION INVERSION

The population inversion required for light amplification constitutes an abnormal distribution of atoms among the various available energy levels. To understand how light amplification can be achieved in a medium, it is necessary to consider what constitutes a "normal" distribution of atomic-energy-state populations. The particular branch of physics that deals with this problem is called statistical mechanics, and the basic principle that describes the distribution is Boltzmann's principle.

Boltzmann's Principle and the Population of Energy Levels

Consider a collection of atoms in a gas* in thermal equilibrium at room temperature. The great majority of the atoms have electron configurations corresponding to their ground state. Some small fraction of atoms are in excited states because of thermal energy that is always available to induce upward transitions. With higher temperatures, more thermal energy is available, and the fraction of atoms in excited states is larger. *Boltzmann's principle* specifies what fraction of atoms are found, on the average, in any particular energy state for any given equilibrium temperature. Stated mathematically,

$$\frac{N_i}{N_0} = e^{-E_i/kT} \tag{3.21}$$

---

* Although lasing systems have been found in every state of matter, we shall refer to a gas as a matter of convenience. What is stated here is applicable to all lasing systems.

where N_i is the number of atoms in the excited state; N_0 is the number of atoms in the ground state; E_i is the energy of the excited state measured relative to the ground state energy; T is the absolute temperature in degrees Kelvin; and $k = 1.38 \times 10^{-23}$ joule/K (Boltzmann's constant). This relationship, known as the *Boltzmann ratio*, is illustrated in Fig. 3.3, where the populations of several energy levels are shown measured relative to the population of the ground state. From this basic relationship, it is possible to obtain several alternative expressions. The ratio of the atomic populations in the gas for two arbitrary energy levels, $E_j > E_i$, for example, is easily shown to be

$$\frac{N_j}{N_i} = \exp\left(\frac{-(E_j - E_i)}{kT}\right) = \exp\left(\frac{-\Delta E_{ji}}{kT}\right). \tag{3.22}$$

Thus, for a specified number of energy levels, a specific temperature, and a known number of atoms in the collection, it is possible to specify the equilibrium number of atoms existing in any particular energy state, E_i.

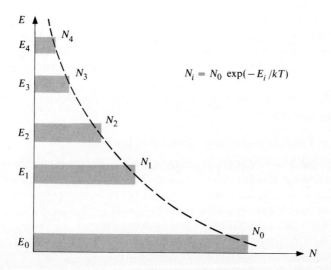

Fig. 3.3 Boltzmann distribution for several energy levels. Dashed line indicates the population of levels if the distribution of energy levels were continuous rather than discrete, as shown here.

To obtain some feel for this distribution, it is useful to consider Eq. (3.22) for two extreme conditions: $\Delta E \ll kT$ and $\Delta E \gg kT$. When $\Delta E \ll kT$, the high-temperature case, the Boltzmann ratio is near unity and the population at energy level E_i is nearly equal to the population at energy level E_j. If E_i should be the ground-level energy, then Eq. (3.22) predicts that atomic states with energies (measured relative to the ground-state energy) small compared to kT are populated to about the same extent as the ground state. If, on the other hand, kT is much smaller than $E_j - E_i$, the low-temperature case, the Boltzmann ratio is quite small and virtually no atoms are found in the higher energy state. A Boltzmann ratio of $1/e$ is often taken as a convenient dividing line between these two extreme situations. Such a ratio occurs when $\Delta E_{ji} = kT$, or, equivalently, when the frequency of the transition satisfies the equation

$$hv_{ij} = kT. \tag{3.23}$$

At room temperature, $T = 300$ K. The corresponding frequency (found by solving for v_{ij} in Eq. (3.23)) has the value of 6×10^{12} Hz. This frequency is in the far infrared part of the spectrum ($\lambda = 50$ μm), and we conclude that for energy gaps with transition frequencies in the near infrared or visible regions, kT is much smaller than ΔE at room temperature, and the higher energy levels are essentially unpopulated; virtually all atoms are in the ground state. This condition is observed in the near infrared and visible absorption spectra of room-temperature gases, where all absorption lines correspond to transitions to some higher-lying energy level from the ground state. Absorption lines corresponding to transitions between higher-lying energy states are not observed. Should the temperature of the gas be increased several thousand degrees, the equilibrium distribution changes to one of greater occupancy of the higher energy levels. Absorption transitions between these higher energy states can now occur, and new lines in the absorption spectrum are observed.

Attainment of Population Inversion: Two-Level Pumping

As T in Eq. (3.22) is increased, N_j approaches N_i but cannot exceed it. Under thermal equilibrium conditions, the population of a higher energy state is never larger than that of a lower energy state. How then is a population inversion obtained in a laser? The answer is that the atoms in the laser medium must be excited—or *pumped*—to a non-thermal-equilibrium distribution through the application of some external source of energy. Once a population inversion is achieved, light amplification by stimulated emission can take place.

There are many different techniques for pumping a collection of atoms to an inverted state. Some of these we shall investigate in Chapter 6 when we consider specific laser systems. Before studying these techniques in detail, however, we can learn a great deal about laser pumping by considering three general schemes: the two-level pumping scheme of the ammonia maser, and the three- and four-level optical pumping schemes typical of many doped insulator lasers.

The particular two-level pumping scheme that we shall consider is not employed for lasers, being in fact unique to the ammonia-beam maser (MASER is the acronym for Microwave Amplification by Stimulated Emission of Radiation). Nevertheless, it serves as an appropriate starting point for a discussion of pumping schemes, illustrating certain general characteristics of pumping. In addition, the ammonia-beam maser has great historical significance, since it was the first man-made device ever to amplify electromagnetic radiation by stimulated emission* and was thus the predecessor of today's lasers. (Lasers were first called optical masers.)

In the ammonia maser, the active medium is a gas of ammonia molecules (NH_3). The maser resonance transition occurs between two quantized energy levels of a particular mode of vibration of the molecule. The energy difference between these two states corresponds to a transition frequency of 24 GHz (24×10^9 Hz), which is in the short-wavelength microwave region of the electromagnetic spectrum. At room temperature, $h\nu/kT \ll 1$ for $\nu = 24$ GHz, and the Boltzmann ratio is near unity. The number of molecules in the higher energy state thus very nearly equals—but does not exceed—the number in the lower state.

The unique aspect of the ammonia-beam pumping technique is that it achieves a population inversion on this transition by effecting a *physical separation* of the higher-energy molecules from the lower-energy molecules. The separation is performed by directing ammonia gas into an electric field gradient (produced by a quadrupole focuser), as shown in Fig. 3.4. Due to differences in the structure of the ammonia molecule in its two energy states, the beam molecules do not all respond to the electric field in the same way. Those molecules in the higher of the two states (designated by ●) experience a net force in the direction of zero electric field, or toward the axis of the focuser. The lower-energy-state molecules (designated by ○), on the other hand, experience a net force away from the axis. These lower-state molecules soon condense on the cool walls of the focuser and are effectively removed from the beam. The molecules remaining in the beam, the great majority of which are in the higher energy state, continue to the end of the focuser, where they enter a microwave cavity that is resonant at the transition frequency. Because the majority of the molecules traversing the cavity are in the ● state, a population inversion exists within the cavity, and any initial electromagnetic disturbance at the transition frequency is amplified by stimulated emission. This amplified disturbance soon builds to a sustained oscillation. The bandwidth of this oscillation is quite small, being governed primarily by the very narrow linewidth of the NH_3 transition (about 6000 Hz). As a consequence, the ammonia maser can be used as a frequency standard in the short-wavelength region of the microwave spectrum.

---

* The first ammonia-beam maser was constructed by Gordon, Zeiger, and Townes at Columbia University in 1954.

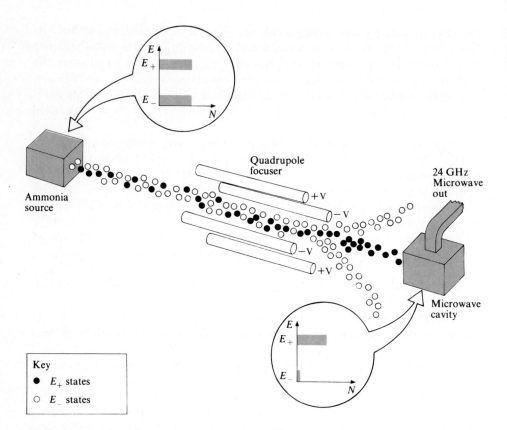

Fig. 3.4 Schematic of the ammonia-beam maser. Because the energy separation of the two states (● and ○) is small compared to the thermal energy of the system ($E_+ - E_- \ll kT$), the energy levels are nearly equally populated (top insert). By passing the atoms through an electric field gradient (quadrupole focuser), the higher-energy-state atoms (●) are directed into a microwave cavity resonant at $v = (E_+ - E_-)/h$. This physical separation creates a population inversion in this two-level system (bottom insert).

It is important to note that no population inversion could be induced on the NH_3 vibrational transition without the mechanism for physically separating the molecules into the two states. One might attempt to pump a majority of molecules into the higher energy state by stimulated absorption, irradiating, for example, a collection of molecules with microwave radiation in a frequency range of about 24 GHz. The resultant absorption transitions would continue, however, only as long as the gas remained absorptive. Through our application of the Einstein relations in the first section of this chapter, however, we showed that the gas is absorptive only as long as the majority of the molecules are in the lower energy state. The harder the molecules are pumped—namely, the more intense the pumping radiation at 24 GHz—the closer the two energy-state populations are driven to equality, until at the limit no more energy is absorbed. A population inversion in a two-level system can never be achieved by optical pumping.

Optical Pumping: Three- and Four-Level Schemes

Optical pumping schemes, where a population inversion is brought about by stimulated absorption, require a process more complicated than the direct pumping of atoms from the ground state to the upper level of a laser transition. The three- and four-level schemes typical of optically-pumped doped insulator lasers over-come the restrictions imposed on a two-level scheme by pumping atoms in the active medium *indirectly* to the upper state of the transition.

The three-level scheme, illustrated in Fig. 3.5, was first proposed by Bloem-bergen at Harvard University in 1956. Initially, the distribution of atomic-state populations obeys Boltzmann's law, as shown in Fig. 3.5(a), and the optical transition between energy state E_1 and the ground state is absorptive. If the col-lection of atoms is intensely illuminated—for example, with a xenon flashlamp—a large number of atoms can be excited through stimulated absorption to the highest energy level, E_2. From there they decay to level E_1, as shown in Fig. 3.5(b). With sufficiently intense pumping, a significant number of ground-state atoms can be pumped to level E_1. A population inversion occurs when the pop-ulation of E_1 exceeds that of the ground state, as shown in Fig. 3.5(c). In order for inversion to be achieved easily, it is necessary for the transition from E_2 to

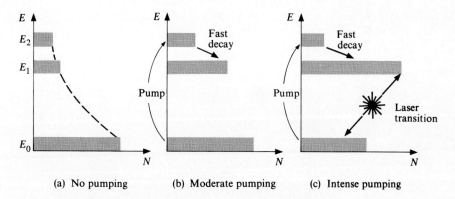

(a) No pumping   (b) Moderate pumping   (c) Intense pumping

Fig. 3.5 Population of energy levels by pumping in a three-level system. (a) Boltzmann distribution of energy states with no pumping. (b) Nonequilibrium distribution with moderate pumping. (c) Population inversion created by intense pumping of the highest energy state followed by decay to the intermediate state.

E_1 to be rapid (i.e., for the $(2 \rightarrow 1)$ transition to have a very short lifetime) and for energy state E_1 to be metastable. If these two conditions are satisfied, ground-state atoms are quickly pumped to level E_1, where they tend to accumulate. The rapid transition from E_2 to E_1 assures that the quantity $N_0 - N_2$ remains large, and the pump transition is highly absorptive.

The ruby laser, discussed in detail in Chapter 6, relies on a three-level pumping scheme, and several other lasers can be characterized by the same scheme. Three-level pumping generally requires very high pump powers, however, because the terminal state of the laser transition is the ground state, and more than one-half of the ground state atoms must be pumped to the upper state to achieve population inversion. The four-level pumping scheme, in contrast, is characterized by greatly reduced pumping requirements. As shown in Fig. 3.6, pump excitation lifts atoms from the ground state to the highest of the four levels involved in the process. From this level, the atoms decay to the metastable state E_2, and the population of this state grows rapidly. If the lifetimes of the $(3 \rightarrow 2)$ and the $(1 \rightarrow 0)$ transitions are short and that of the $(2 \rightarrow 1)$ transition is long, a population inversion on the $(2 \rightarrow 1)$ transition can be achieved and maintained with only moderate pumping. The populations of energy levels E_3 and E_1 remain essentially unchanged with the onset of pumping, and only a small number of atoms need

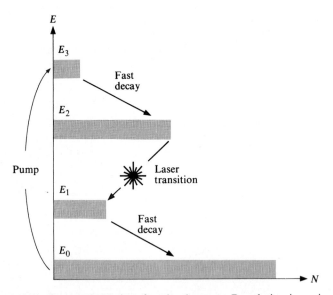

Fig. 3.6 Population of energy levels in a four-level system. Population inversion is achieved by pumping to a high energy state (E_3), followed by fast decay to a lower energy state (E_2). Because the ground state is not the lower level of the population inversion, the ground-state population may be large, yet the pumping needed to achieve inversion need not be intense.

be added to the population of energy level E_2 to make N_2 greater than N_1. In Chapter 4, we discuss a four-level system in greater detail.

The detailed mechanisms employed in pumping lasers can be quite complicated and varied. In addition to the optical pumping schemes just described, there are pumping techniques that rely on electrical discharge, the combustion of fuel, or pressure-temperature cycling. Frequently these schemes involve many more than the three or four energy levels discussed here, and complicated interactions between energy levels occur. Nevertheless, the simple optical pumping schemes presented here provide remarkably good models for pumping in the majority of real cases.

3.3 OPTICAL FEEDBACK: THE LASER RESONATOR

The most direct application of the amplification-by-stimulated-emission principle is incorporated in a *single-pass light amplifier*, a device that takes as its input a collimated beam of narrowband light, typically from a laser, and amplifies it. Light amplifiers are regularly employed to increase the output of ruby lasers and neodymium-doped glass lasers.

The Optical Resonator

Direct amplification of light in the manner described above is practical only for a few select materials. This is because the amount of amplification provided by most active media (the collections of atoms, ions, or molecules participating in the stimulated-emission process) is measured in fractions of a percent per centimeter of light travel—a level much too low to be useful in a single-pass device of reasonable length. In most lasers, this limitation is circumvented by the use of mirrors that direct the light beam back and forth through the active medium many times. With such an arrangement, the effective length of the amplifying medium becomes many times the actual length of the laser. A laser, for example, that has a 100-percent reflective mirror at one end and a 98-percent reflective mirror at the other end has an effective length roughly 50 times the actual separation distance between mirrors. Although single-pass amplification may be quite small for the inverted medium, the total amplification provided by multiple passes can be substantial.

There are some lasing systems in which it is not possible or practical to use mirrors because of the pumping conditions or the lifetime of the population inversion. Normally these systems have sufficient amplification to produce a partially coherent output whose frequency spread is considerably narrower than their spontaneous emission linewidth. These systems, which operate with an inverted medium but without mirrors for multiple reflections, are referred to as *superradiant*. While it would probably be useful to refer to these systems as light amplifiers rather than lasers, the terms are used interchangeably.

Round-Trip Power Gain and the Threshold Condition

Those lasers with highly reflective mirrors are not really amplifiers in the usual sense, but are oscillators. The mirrors introduce feedback (see box), forming an *optical resonator* or *optical resonant cavity* that supports continuous electromagnetic oscillations at the laser transition frequency. Initially, a small disturbance at the resonance frequency, provided by spontaneous emission within the cavity, undergoes amplification. However, this amplification continues (at least in continuous-output lasers) only until some *steady-state* level of oscillation is reached.* At that point, growth of the waves within the cavity ceases; all additional energy supplied by the stimulated-emission process then serves to maintain the laser output. A necessary requirement for the oscillations to be established is an inversion of the populations of the energy levels of the laser transition. It is not sufficient, however, to require simply that the population of the upper state be greater than that of the lower state. Various losses, including the loss of power in the form of the laser output, must be accounted for. To overcome these losses, a minimum gain coefficient, the *threshold gain*, β_{th}, is required to initiate and sustain laser oscillations. This threshold gain, in turn, through Eq. (3.20), specifies the minimum population inversion required.

We can determine the minimum gain necessary for operation of a laser by considering the increase of irradiance undergone by a beam of light at the resonance frequency in traveling a round-trip path within the laser. To simplify the analysis, we assume that the active medium completely fills the region between the mirrors and that the pumping excitation is uniform. In traveling from mirror

---

FEEDBACK

The phenomenon of feedback in public address systems is familiar to most people. In some ways it parallels the feedback processes in a laser. If some sound from the amplifier is picked up by the microphone, it is further amplified. Because there is a necessary delay in the time between sound output by the amplifier and its pickup by the microphone, a periodic variation in the sound amplitude occurs. The amplifying system becomes an oscillator emitting an unnerving, high-pitched squeal. Likewise in the laser, the gain of the amplifier saturates and steady-state oscillations are sustained.

For the present we use the term "feedback" rather loosely, implying only that a part of the light within the cavity formed by the mirrors is "fed back" to be further amplified. It should be noted, however, that a precise feedback-based analysis, similar to that employed in the analysis of certain types of amplifiers (known as *regenerative-feedback amplifiers*), is often used in the study of lasers. See, for example, Section 5.5 of Ref. 3.5.

---

* Perhaps the acronym should come from Light *Oscillation* by Stimulated Emission of Radiation. But then nobody wants to be associated with a LOSER.

M_1 to mirror M_2, the beam irradiance increases from its initial value, I_0, to a value I given by

$$I = I_0 e^{(\beta - \alpha_l)L} \tag{3.24}$$

where L is the separation of the mirrors, β is the small-signal gain coefficient, and α_l is the distributed loss per unit distance due to scattering and possible absorption in nonactive constituents of the laser medium. After reflection at mirror M_2, the relative beam irradiance is given by $R_2 \exp[(\beta - \alpha_l)L]$, where R_2 (the reflectivity of mirror M_2) is the ratio of reflected to incident light irradiance. After a return trip through the active medium and a reflection at mirror M_1 (which has reflectivity R_1), the final round-trip relative increase in beam irradiance is given by

$$G = R_1 R_2 \exp[2(\beta - \alpha_l)L]$$

$$= \frac{\text{Beam irradiance at end of round-trip}}{\text{Beam irradiance at start of round-trip}}. \tag{3.25}$$

G, which is the *net round-trip power gain* of the laser (related exponentially, it should be noted, to the small-signal gain, β), quantifies the buildup of oscillations within the laser as a function of β and therefore, by Eq. (3.20), as a function of the population difference. If G is greater than unity, a disturbance at the resonance frequency of the laser experiences a net round-trip growth in magnitude and the cavity oscillations increase. If, on the other hand, G is less than unity, oscillations within the cavity die out. We can therefore write the *threshold condition for laser oscillation* as

$$G = R_1 R_2 \exp[2(\beta - \alpha_l)L] = 1. \tag{3.26}$$

In a laser designed for continuous output, the small-signal gain, which may have increased momentarily to well above the threshold level with the onset of pumping, drops down to β_{th}. To see why this is so, consider that a net round-trip power gain greater than unity causes the cavity energy density to increase and that a power gain less than unity causes the energy density to decrease. Only when $G = 1$ for a period of time does the cavity energy (and therefore the laser output power) settle down to a steady-state value. Therefore, the steady-state value of the small-signal gain must equal the threshold value, β_{th}. This pinning of the steady-state gain at the threshold value is referred to as *gain saturation*.

The small-signal gain necessary to support such steady-state operation depends both on the constituents of the laser medium (through α_l) and on the laser construction (through R_1, R_2, and L). We can obtain an expression for β_{th} in terms of these parameters by taking the logarithm of both sides of Eq. (3.26), with the result

$$\beta_{th} = \alpha_l + \frac{1}{2L} \ln\left(\frac{1}{R_1 R_2}\right) = \alpha_l + \alpha_0 \tag{3.27}$$

a form that emphasizes the similar nature of the two terms as losses: α_l represents losses of cavity radiation resulting from absorption and scattering in the medium; α_0 represents the radiation lost to the cavity in the form of useful output. Thus, when we regard the useful output as a loss, we can state the following condition for steady-state laser oscillation: *The gain is equal to the sum of the losses in the laser.*

As expressed in Eq. (3.20), β can assume a wide range of values, depending not only on $\Delta\mathcal{N}$ but also on intrinsic properties of the active medium. A laser with a high-gain active medium can be made to lase comparatively easily. Alignment of the mirrors is not very critical; some dust and imperfections in the reflective mirror surfaces can be tolerated. In short, losses (including output) can be greater because a higher threshold gain is allowable. With low-gain mediums, on the other hand, reflectivities must be very high, losses made as low as possible, and mirror surfaces critically aligned before lasing can be initiated.

High, steady-state gain should not be confused with high efficiency. The efficiency of a laser is determined by taking the ratio of output laser light power to input pumping power. It depends upon how well the pump power is channeled

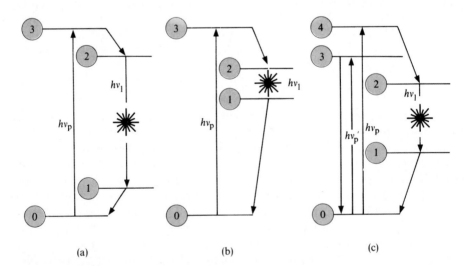

(a)   (b)   (c)

Fig. 3.7 Comparison of efficiencies of lasers based on a four-level system. (a) This scheme would be highly efficient because most of the pump energy, $h\nu_p$, is converted to laser energy, $h\nu_1$, and all of the pump energy is used to populate the upper laser state. The only losses are due to the nonlasing transitions $(3 \rightarrow 2)$ and $(1 \rightarrow 0)$. (b) This scheme is less efficient than that in (a) because the laser transition energy is a small fraction of the pump transition energy. (c) This scheme is also less efficient than that in (a), since much of the pump energy is channeled into nonlasing transitions.

into inverting the laser transition populations and on the probabilities for different kinds of transitions from the upper energy state of that transition. Several examples are given in Fig. 3.7. Certain lasers, the CO_2 laser being an excellent example, are characterized by both high efficiency (5 to 35 percent) and high small-signal gain. Other lasers, the argon laser for example, though having high gain, are characterized by very low efficiency (0.001 to 0.1 percent). Still other lasers are doubly cursed, having both low gain and low efficiency.

Confinement of the Beam within the Resonator: The Stability Condition

Thus far in our discussion of the optical resonator, we have tacitly assumed that radiation within the laser cavity propagates back and forth between plane-parallel mirrors in the form of a well-collimated beam. Because of diffraction effects, however, a perfectly collimated beam cannot be maintained with mirrors of finite extent, and a fraction of the internal cavity radiation spills out around the edges of the mirrors. This unconfined radiation no longer contributes to the stimulated emission process in the cavity and represents a loss that must be balanced by a higher small-signal gain (and hence harder pumping) if laser action is to be achieved. For many mediums and pumping schemes, such additional losses mean the difference between sufficient and insufficient gain to reach threshold. Only by designing the laser resonator to assure almost total confinement of the beam within the cavity have scientists been able to obtain outputs from many laser mediums.

In general, the problem of excessive diffraction losses is overcome by curving the mirrors slightly inward toward the cavity, that is, by making them slightly spherical and concave inward. A detailed description of the effect of this mirror curvature on laser operation would require a rigorous application of diffraction theory. It is possible, however, through an application of much simpler ray-tracing techniques, to determine those combinations of spherical mirrors that correspond to high-loss configurations and those that correspond to low-loss configurations. The actual analysis is still somewhat tedious and is not presented here. We describe the essential steps of the analysis, however, and discuss the final results.

In Fig. 3.8 is illustrated a pair of spherical mirrors, each positioned with its center of curvature on a common line, the optic axis of the resonator. The mirrors are separated by a distance L; r_1 is the radius of curvature of mirror M_1, r_2 that of mirror M_2. By convention, the radius of curvature of a mirror is taken to be positive if the center of curvature of the mirror lies in the direction of the laser cavity; otherwise r is taken to be negative. (In the figure, both radii are positive by this convention.) The ray-tracing technique requires that we consider a ray of light within the cavity, initially very close to the optic axis and inclined at a very small angle with respect to it (such a ray is designated a *paraxial ray*), and that we observe its behavior as it is reflected back and forth between the mirrors.

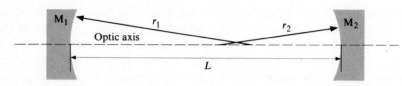

Fig. 3.8 Optical resonator consisting of two spherical mirrors. Radii are defined as having positive values if the mirrors are concave.

If after a large number of reflections the ray is observed to diverge from the resonator—that is, the distance of the ray from the optic axis is found to grow as the number of reflections becomes large—we conclude that the particular resonator configuration being analyzed is characterized by high losses. If, on the other hand, the ray is found to remain close to the optic axis, we conclude that the resonator has a low-loss configuration.

By performing such an analysis for an arbitrary mirror separation L and arbitrary mirror curvatures r_1 and r_2, one can show that a single condition, called the *stability condition*, is satisfied by a low-loss resonator configuration.* This condition is generally expressed in terms of two dimensionless quantities, the *g-parameters* of the resonator, defined by the equations

$$g_1 = 1 - \frac{L}{r_1} \quad \text{and} \quad g_2 = 1 - \frac{L}{r_2} \tag{3.28}$$

for mirror M_1 and mirror M_2, respectively. With these definitions, the stability condition has the simple form

$$0 < g_1 g_2 < 1. \tag{3.29}$$

As long as this relationship is satisfied by the resonator, a paraxial ray continues to remain close to the optic axis, even after many reflections, and the mirror configuration is termed *stable*. If, on the other hand,

$$g_1 g_2 < 0 \quad \text{or} \quad g_1 g_2 > 1 \tag{3.30}$$

the resonator is described as being *unstable*, and upon multiple reflections a ray initially paraxial in the cavity diverges from the axis. In the cases where the product $g_1 g_2$ equals zero or unity, the laser is on the boundary between stability and instability and is termed *marginally stable*. In Fig. 3.9, we have illustrated some important configurations and indicated their stability.

---

* See Chapter 8 of Ref. 3.5 for the actual analysis.

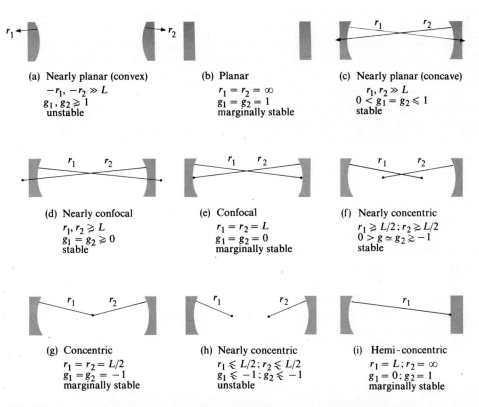

(a) Nearly planar (convex)
$-r_1, -r_2 \gg L$
$g_1, g_2 \gtrsim 1$
unstable

(b) Planar
$r_1 = r_2 = \infty$
$g_1 = g_2 = 1$
marginally stable

(c) Nearly planar (concave)
$r_1, r_2 \gg L$
$0 < g_1 = g_2 \leqslant 1$
stable

(d) Nearly confocal
$r_1, r_2 \gtrsim L$
$g_1 = g_2 \gtrsim 0$
stable

(e) Confocal
$r_1 = r_2 = L$
$g_1 = g_2 = 0$
marginally stable

(f) Nearly concentric
$r_1 \gtrsim L/2; r_2 \gtrsim L/2$
$0 > g \simeq g_2 \gtrsim -1$
stable

(g) Concentric
$r_1 = r_2 = L/2$
$g_1 = g_2 = -1$
marginally stable

(h) Nearly concentric
$r_1 \lesssim L/2; r_2 \lesssim L/2$
$g_1 \lesssim -1; g_2 \lesssim -1$
unstable

(i) Hemi-concentric
$r_1 = L; r_2 = \infty$
$g_1 = 0; g_2 = 1$
marginally stable

Fig. 3.9 Laser cavity mirror configurations. Stability for each of these configurations is indicated.

It should be noted that although the ray-tracing techniques employed cannot predict many of the details of optical-resonator behavior that are revealed by a more rigorous analysis based upon diffraction theory, the stability condition expressed in Eq. (3.29) is always applicable.

The most common laser resonator is a nearly planar stable configuration or an equivalent configuration with one long radius-of-curvature mirror and one planar mirror. Such a configuration is relatively easy to align. It is also efficient, in the sense that a reasonably large fraction of the resonator volume is utilized in the amplification process.

Even though a particular resonator configuration is classified as unstable, it may still be quite useful. Indeed, unstable optical resonators have several very attractive features. First, they can be highly efficient with respect to utilization of the active medium, even with very short resonators. In Chapter 2, we noted that in the typical laser resonator, the beam tapers down to a narrow waist. Only that portion of the active medium near the optic axis contributes much to the amplification of the beam. By way of contrast, the entire volume of the active medium in an unstable laser resonator can contribute strongly to the amplification process. A second attractive feature of unstable resonators is their suitability for adjustable output coupling. Simply by changing the spacing between the resonator mirrors, one can adjust the output coupling over a wide range of values. In Fig. 3.10, a typical stable laser resonator is compared with a typical unstable resonator. The output of the unstable resonator passes around one of the resonator mirrors and, close to the laser, the laser beam has an annular cross section. This characteristic cross section does not significantly reduce the usefulness of the output of an unstable resonator. It is still possible, for example, to focus the output to a spot that is most intense at the center, very much like the focused output of a laser with a Gaussian beam. Because the unstable resonator exhibits large losses (indeed, from Fig. 3.10 it should be clear that these losses constitute the output of the laser!), it can be used only with an active medium that has a high small-signal gain, e.g., carbon dioxide or carbon monoxide. Under those circumstances, however, an unstable resonator provides the best practical method for obtaining nearly complete energy extraction from the laser medium in an output beam of high optical quality.

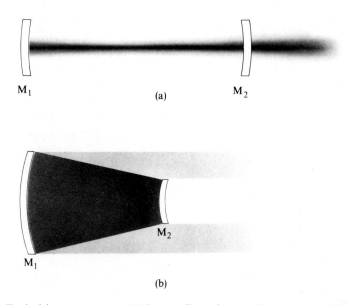

Fig. 3.10 Typical laser resonators: (a) long-radius mirror stable resonator; (b) confocal, or "telescopic," unstable resonator.

The Basic Laser: Simple Version

In this chapter, we have presented a first look at lasers. At this point, our picture of a laser is that of a highly collimated source of monochromatic light whose output is steady, clear, and true. It is a close approximation to the popular science writer's picture of a laser. But our picture is still far from complete. The laser is much more complex and interesting than the device presented in this chapter. In the following chapters, we shall investigate various aspects of the laser output and construction that serve to distinguish among types of lasers and among applications for which they can be employed.

PROBLEMS

3.1   Express the outputs of the following lasers in terms of (a) units of frequency (hertz) and (b) units of energy (joules/photon).

Laser	Wavelength of output
Nitrogen	337.1 nm
Argon ion	488 nm, 514.5 nm
Helium-neon	632.8 nm
Neodymium	1.06 μm
Carbon dioxide	9.6 μm, 10.6 μm

3.2   The first few rows of the periodic table of the elements can be explained by a simple application of the Pauli exclusion principle. List the quantum numbers for each electron of the sodium atom in a magnetic field.

3.3   List the quantum numbers for the electrons in the ground state of the neon atom. What are the quantum numbers of the next state to be filled in this scheme? If a single electron were excited by the minimum amount of energy possible, what would then be the configuration of the atom?

3.4   If the spontaneous emission rate, A_{10}, is 10^6/sec for an x-ray wavelength transition ($\lambda = 100$ nm),

a)   what is B_{10} for this transition?
b)   what must be the spectral irradiance (at 100 nm) within the cavity to cause stimulated emission three times greater than the spontaneous emission rate?

3.5   If $B_{10} = 10^{19}$ m³/watt-sec³, what is A_{10} and the corresponding lifetime τ_{10} (its reciprocal) for

a) $\lambda = 600$ nm (visible)?
b) $\lambda = 6$ μm (infrared)?
c) $\lambda = 60$ nm (ultraviolet)?
d) $\lambda = 0.6$ nm (x-ray)?

How does this compare to a stimulated transition rate if $I = 10$ watts/mm²?

3.6   If 1 percent of the light transmitted into a cell is absorbed per millimeter, what fraction is transmitted by a cell of 10 cm length? Calculate α.

3.7   If the irradiance of light doubles on a pass through a laser amplifier of one meter active length, calculate β, assuming no losses. If there were only a 6-percent increase in irradiance in that length, what would β be?

3.8   If 10 percent of the total population is considered an "appreciable fraction," (a) what temperature would be needed to obtain such a population if the absorption to the state were in the visible region of the spectrum ($\lambda = 500$ nm)? (b) For the ammonia maser transition, how cold would the material (gas, liquid, or solid) have to be to have only an "appreciable fraction" of atoms in the excited state of the transition?

3.9   An atom has two energy levels with a transition wavelength $\lambda = 694.3$ nm. What percentage of the atoms are in the upper state at room temperature ($T = 300$ K)? Assume that all the atoms are in one or the other of these two energy levels.

3.10   The stability condition (Eq. 3.29) can be represented graphically.

a)   In a Cartesian coordinate system, let g_1 be the abscissa and g_2 be the ordinate. Plot the boundary curves and shade those regions associated with values g_1, g_2 satisfying $0 < g_1 g_2 < 1$, that is, for which the resonator is stable. The resulting diagram is the so-called *stability diagram*.

b)   On your diagram, locate the points (g_1, g_2) corresponding to the resonators shown in Fig. 3.9. Label each point with its corresponding letter from the figure.

3.11   Given two mirrors with radii of curvature $r_1 = +50$ cm and $r_2 = +100$ cm, calculate (a) the separations for which they would form a marginally stable configuration, (b) the range of separations for which they would be stable, and (c) the range of separations for which they would be unstable.

REFERENCES

3.1   Editors of *Scientific American* (1969), *Lasers and Light*. San Francisco: W. H. Freeman. Reprints of *Scientific American* articles with introductions to major sections by A. L. Schawlow. Nonmathematical readings. The laser-related articles occupy the final third of the book. Some of the sections are dated.

3.2   O. S. Heavens (1971), *Lasers*. New York: Scribner's. Intended for readers with a limited scientific and mathematical background.

3.3   B. A. Lengyel (1971), *Lasers* (2nd edition). New York: Wiley.

3.4   A. Maitland and M. H. Dunn (1970), *Laser Physics*. New York: American Elsevier.

3.5   A. E. Siegman (1971), *Introduction to Lasers and Masers*. New York: McGraw-Hill. Nonquantum mechanical treatment that leans heavily on electrical engineering concepts.

3.6   O. Svelto (1976), *Principles of Lasers*. New York: Plenum. Up-to-date discussion of lasers, requiring a background in elementary quantum mechanics.

4

The Laser Output

As we have seen in the previous chapter, most lasers meet the three basic requirements: an active medium, a population inversion, and a feedback mechanism. In this chapter, we investigate how these requirements determine the temporal, spatial, and spectral characteristics of the laser output. In the course of this investigation, we shall consider in detail the atomic lineshape associated with a laser transition, the resonance characteristics of the optical cavity formed by the laser mirrors, and the amplification process. In the next chapter, we shall discuss how the temporal, spatial, and spectral characteristics can be controlled and modified through proper design of the laser system.

4.1 ABSORPTION AND EMISSION LINESHAPE BROADENING

In deriving the relation for the small signal gain in the previous chapter, we assumed that all the radiation present was concentrated at the resonance frequency, v_{ij}. In fact, absorption and emission do not take place at a single frequency, but over a range of frequencies spanning the atomic linewidth. The absorption and small-signal gain coefficients, α and β, are really functions of frequency, $\alpha(v)$ and $\beta(v)$. In order to account for this frequency dependence of absorption and emission by a particular atomic transition, we replace the expression for α (Eq. 3.18) by the more general relation

$$\alpha(v) = B_{ij}(\mathcal{N}_i - \mathcal{N}_j)\frac{hv_{ij}}{4\pi c}g(v) \tag{4.1}$$

where $g(v)$, called the *lineshape function*, contains all the frequency dependence of α. To be consistent with our earlier definitions, $g(v)$ is equal to unity at the resonance frequency, i.e., $g(v_{ij}) = 1$. Our old α is then seen to equal the new $\alpha(v_{ij})$, i.e., the absorption coefficient at line center.

Consistent with Eq. (3.20), β has a similar dependence on v:

$$\beta(v) = -\alpha(v) \tag{4.2}$$

or

$$\beta(v) = B_{ij}(\mathcal{N}_j - \mathcal{N}_i)\frac{hv_{ij}}{c}g(v)$$

for $(\mathcal{N}_j - \mathcal{N}_i) > 0$, $E_j > E_i$. In this section, we investigate various mechanisms, collectively referred to as *lineshape broadening mechanisms*, that account for the frequency dependence of α and β and for the corresponding finite linewidths associated with emissive and absorptive transitions.

The width of an atomic transition lineshape can be expressed in several ways. In Fig. 4.1(a), the linewidth of an emission line has been chosen as the distance between points halfway down the lineshape curve. Such a definition, based on the *full width at half maximum* power (FWHM), is generally convenient and is the most common. There are circumstances, however, when an alternative definition may be chosen. Consider, for example, the emission radiance versus frequency curve for the 632.8-nm neon line, characteristic of the helium-neon laser, illustrated in Fig. 4.1(b). For reasons to be explained later, this particular lineshape is Gaussian, falling off as

$$e^{-(v-v_0)^2/2\sigma^2}$$

where v_0 is the center frequency, or *resonance frequency*. A particularly appropriate choice for the linewidth in this case is twice the standard deviation of the

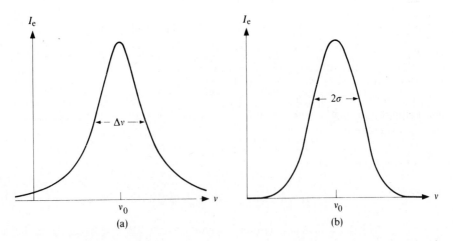

Fig. 4.1 (a) The width of an atomic resonance transition lineshape is frequently expressed as the distance between the half-maximum power points. (b) When the resonance curve is Gaussian in shape, as is often the case with gas discharge tubes, the linewidth is frequently expressed as twice the standard deviation (σ) of the Gaussian curve.

curve, σ, as shown in the figure.* The choice of units (hertz, angstroms, nanometers, etc.) used to express these linewidths depend on whether a frequency band or a wavelength spread is to be described.

There is a variety of mechanisms that contribute to the broadening of the spectral lineshape of emitting and absorbing atoms. We consider here the three most important: Doppler broadening, collision broadening, and lifetime broadening.

Doppler Broadening

When an automobile zips past you with its horn blaring, the pitch of the horn is higher as the auto approaches you and lower as the auto recedes. This phenomenon, called the *Doppler effect*, occurs because the velocity of the source causes either a greater or lesser number of oscillations to reach the ear in a given time period, thus raising or lowering the perceived frequency of the source.

The predominant broadening mechanism of the active media of most gas lasers, known as *Doppler broadening*, is a direct result of the Doppler effect. Consider an analogy. We provide a skater at an ice rink with a 1000-Hz tin whistle. If the skater is standing still on the ice, we hear the 1000-Hz note when he or she blows the whistle. If this person skates toward us with a velocity v_z, the number of vibrations of the air on our eardrum during each second is increased from v_0 (1000 Hz) to

$$v' = v_0 \left(1 - \frac{v_z}{v_s}\right)^{-1} \tag{4.3}$$

where v_s is the velocity of sound in air (330 m/sec). If the person skates away from us at the same speed, the frequency we hear is decreased to

$$v' = v_0 \left(1 + \frac{v_z}{v_s}\right)^{-1}. \tag{4.4}$$

Suppose now that a large number of the skater's friends come to the rink and are also issued 1000 Hz tin whistles. If everyone toots his whistle whenever he feels like it and skates in any direction and at any speed that he pleases, we hear a large number of different frequencies, some greater than v_0, some less than v_0, and some at v_0, depending on whether the skater has a relative velocity component toward us, away from us, or none at all. The overall effect will be a band of frequencies centered around 1000 Hz, whose distribution depends on the velocities of the skaters on the rink relative to us.

This effect applies to a gas of atoms emitting at an optical frequency v_{ij}; we simply replace the speed of sound, v_s, with the speed of light, c, in the above equations and let v_z be the velocity of the atoms relative to the observer. Then

$$v' = v_{ij} \left(1 \pm \frac{v_z}{c}\right)^{-1} \tag{4.5}$$

---

* The standard deviation has the same units as the variable (in the above case, hertz or cycles per second) and specifies the breadth of the line. At the frequency v at which $v - v_0 = \sigma$, the value of the Gaussian curve is $e^{-1/2}$, or 61 percent of the value at the peak ($v = v_0$).

where the negative sign is for approaching relative motion of atoms and observer, and the positive sign is for receding relative motion. Just as in the case of the skaters, different atoms in the gas, having different components of velocity in the direction of the observer, have different apparent resonant frequencies. The cumulative result is a smearing or broadening of the emission lineshape. It is convenient to view the broadened lineshape as being made up of a large number of "spectral packets" centered at different frequencies throughout the lineshape. The individual spectral packet at a particular frequency v is the contribution of that fraction of atoms whose Doppler-shifted resonance frequencies lie within a small range of frequencies about v. Many such packets added together produce the smooth Doppler-broadened lineshape.

Under conditions of equilibrium, the distribution of atom velocities in a gas at temperature T is given by a Gaussian (Maxwellian) distribution:

$$\frac{\Delta N(v_z)}{N} = \sqrt{\frac{M}{2\pi kT}} \exp\left(-\frac{Mv_z^2}{2kT}\right)\Delta v_z \tag{4.6}$$

where $\Delta N(v_z)/N$ is the fraction of atoms in the collection with velocity lying in the narrow range between v_z and $(v_z + \Delta v_z)$, M is the atomic mass, and k is the Boltzmann constant. When each spectral packet is weighted by the fraction of atoms contributing to it, the resultant lineshape is itself Gaussian, as illustrated in Fig. 4.2, if the spread in velocities is sufficiently large compared to the width of the individual spectral packets. The width of the individual spectral packets is determined by other broadening mechanisms. By combining the Maxwellian distribution, Eq. (4.6), with Eq. (4.5) for the Doppler shift, the actual Gaussian lineshape observed under these circumstances can be shown to have the form

$$g_G(v) = \frac{I_v(v)}{I_v(v_{ij})} = \exp\left(\frac{-Mc^2}{2kT}\frac{(v - v_{ij})^2}{v_{ij}^2}\right). \tag{4.7}$$

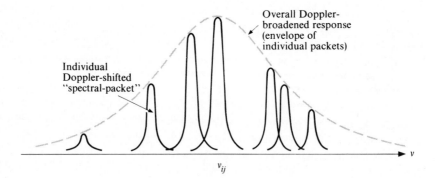

Fig. 4.2 The Doppler-broadened lineshape is made up of the superposition of a large number of individual "spectral packets," which are the contributions from different groups of atoms with different Doppler velocities.

The linewidth for this Doppler-broadened distribution can be expressed either as twice the standard deviation for the Gaussian curve,

$$\Delta v_G = 2 \sqrt{\frac{kT}{Mc^2}} v_{ij} \tag{4.8}$$

or as the distance between the half-maximum-value points,

$$\Delta v_{HM} = \sqrt{(8 \ln 2) \frac{kT}{Mc^2}} v_{ij} = 2.35 \sqrt{\frac{kT}{Mc^2}} v_{ij}. \tag{4.9}$$

In either case, we see that the linewidth increases with increasing temperature and decreases with increasing atomic mass. As an example, the half-maximum linewidth for the 632.8-nm HeNe laser transition, assuming a discharge temperature of about 400 K and a neon atomic mass of 20, is about 1500 MHz.

Collision Broadening

Doppler broadening is the predominant broadening mechanism in most gas lasers emitting in the visible part of the spectrum. The temperature is elevated, the resonance frequency is high, and the atomic mass is relatively low—all these conditions contribute to Doppler spreading. Molecular gas lasers are another story, however. These lasers emit in the infrared part of the spectrum, with resonance frequencies and, therefore, Doppler linewidths substantially smaller than those of their visible-emission counterparts. For such lasers, another mechanism known as *collision broadening* becomes important.

In a gas at elevated temperatures, such as in the active discharge region of a gas laser, collisions occur between the atoms of the active medium and other atoms in the gas. If an atom is emitting a photon when such a collision occurs, the phase of the radiated wavetrain is suddenly altered. Over the period during which it is being emitted, we can approximately describe the wave by the function $\cos[\omega t + \varphi(t)]$, where $\varphi(t)$ is constant most of the time, but changes discontinuously with each collision. These random, discontinuous changes in phase serve to broaden the bandwidth of the wavetrain. When a great many such atoms are in a collection, each atom experiencing random dephasing collisions, the resultant emission lineshape has the form

$$g_L(v) = \frac{I_v(v)}{I_v(v_{ij})} = \frac{1}{1 + \left(\dfrac{4\pi(v - v_{ij})}{\dfrac{2}{T_2}} \right)^2} \tag{4.10}$$

where T_2 is a characteristic time, called the *collision time* or *dephasing time*, that depends on the temperature and pressure of the gas and on the mass of the atoms. This spectral distribution, called a *Lorentzian lineshape* after physicist H. A. Lorentz, is illustrated in Fig. 4.3 and in Fig. 4.1(a), although not labeled as such. Note that the Lorentzian lineshape tends toward zero less rapidly in the wings (Fig. 4.1a) than does the Gaussian lineshape (Fig. 4.1b).

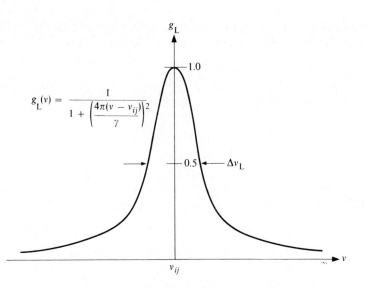

Fig. 4.3 The Lorentzian lineshape can represent either the intensity emission or absorption curve associated with a particular atomic transition. It is characteristic of lifetime-broadening mechanisms. The function is normalized to a value of unity at the central frequency, v_{ij}.

Collision broadening is not unique to gases; it is, in fact, a principal broadening mechanism in doped insulator lasers, for example, ruby and neodymium-doped glass lasers. In such lasers, the atoms or ions of the active medium are, for all practical purposes, locked in place in a crystalline or amorphous solid, and the collisions can no longer be visualized as the impacting of two particles. Nonetheless, there are close interactions between the atoms or ions of the active medium and the neighboring atoms or ions in the surrounding solid. The net result of these interactions is essentially the same as the result of the collisions that occur in the gas.

In some cases, neither collision nor Doppler broadening is predominant. In contrast to the helium-neon laser, in which the broadening is due primarily to the Doppler effect, the CO_2 laser, with larger molecular mass and smaller resonance frequency, is affected roughly equally by Doppler broadening and collision broadening. Under these circumstances, neither the Gaussian function nor the Lorentzian function provides a valid expression of the emission lineshape, and a more complicated combination of the two must be used.*

---

* It is possible through a clever use of tunable lasers and beam geometry to study collision effects in gases even though Doppler broadening is much greater than collision broadening. This new field of *Doppler-free spectroscopy* is one of many surprising applications of lasers that have arisen in the past decade.

Natural Radiative Lifetime Broadening

Suppose that a solid, in which there is no appreciable Doppler broadening, is cooled to reduce all collision broadening to an insignificant level. Can we then expect the linewidth of the emission to be infinitely narrow, the radiation monochromatic? No. There is another mechanism known as *natural radiative lifetime broadening* that imposes a fundamental limit on the bandwidth of the radiation. The oscillatory wavetrains emitted by an atom, no matter how long in duration, are not infinitely long and therefore not truly sinusoidal. As the waves are not strictly sinusoidal, they are not monochromatic, but have a spectrum that occupies a small band about v_{ij}. We can determine the spectrum of these waves by analyzing a simple model for an isolated atom, as shown in Fig. 4.4.

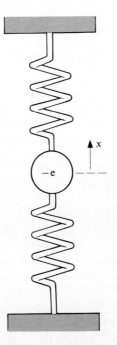

Fig. 4.4 Simple mechanical model for the classical electron oscillator. Electromagnetic radiation, resulting from the acceleration of the charge, provides a natural damping mechanism.

A mass with an electronic charge $-e$ is bound by springs and oscillates with a resonance frequency v_{ij}. The electron is displaced a distance x from its equilibrium position and released. The subsequent oscillations cause accelerations of the electron, and it is known from electromagnetic theory that when an electron is accelerated, electromagnetic energy is radiated. This radiation of energy provides an inherent damping mechanism that leads to an exponential decay of the amplitude of the oscillations.

In the absence of damping, the differential equation describing the motion of the oscillator can be written as

$$\ddot{x}(t) + (2\pi v_{ij})^2 x(t) = 0 \tag{4.11}$$

with solution

$$x(t) = x(0) \cos 2\pi v_{ij} t. \tag{4.12}$$

Radiative energy dissipation can be taken into account by introducing a term in the equation of motion proportional to $\dot{x}(t)$, the oscillator velocity,

$$\ddot{x}(t) + \gamma \dot{x}(t) + (2\pi v_{ij})^2 x(t) = 0. \tag{4.13}$$

In the case of interest, $\gamma \ll v_{ij}$, this equation has as its solution an exponentially damped sinusoid,

$$x(t) \cong x(0) e^{-1/2(\gamma t)} \cos 2\pi v_{ij} t. \tag{4.14}$$

The energy stored in the oscillator, $w(t)$, is calculated by squaring the amplitude, $x(t)$, averaging over several oscillations, and introducing a constant of proportionality. The resultant expression for $w(t)$ is

$$w(t) = w_0 e^{-\gamma t}. \tag{4.15}$$

The energy decays exponentially, as illustrated in Fig. 4.5, with a time constant $1/\gamma$. γ is appropriately referred to as the *energy damping rate*. The simple model above provides a remarkably accurate description of an isolated atom undergoing a transition between two energy states. Using classical electromagnetic theory, γ can be calculated and found to agree with the experimentally observed damping rate.

The spectrum associated with a true sinusoid is a single, infinitesimally narrow line. With any deviation from a pure sinusoid, broadening of the spectrum occurs. For the special case of an exponentially damped sinusoid, it can be shown that the emission intensity spectrum has the same Lorentzian form as that given in Eq. (4.10) for collision broadening,

$$g_L(v) = \frac{I_v(v)}{I_v(v_{ij})} = \frac{1}{1 + \left(\dfrac{4\pi(v - v_{ij})}{\gamma}\right)^2}. \tag{4.16}$$

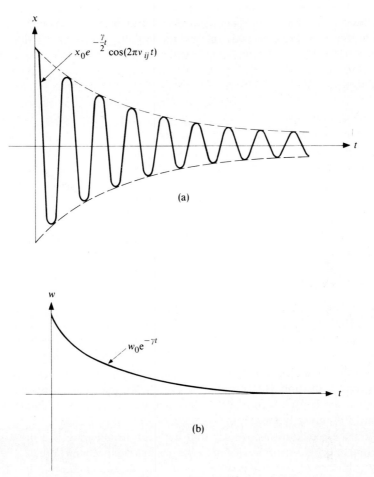

Fig. 4.5 Natural response of the electron oscillator. (a) The displacement (x) follows an exponentially damped sinusoid. (b) The oscillator energy decays with an exponential rate, γ. For actual atomic resonances, v_{ij} is much larger than γ, and many oscillations take place before an appreciable fraction of the stored energy is radiated.

The damping rate is directly related to the half-intensity linewidth, Δv_{L}, of the curve. Narrow linewidths are associated with low damping, or small values of γ, and broad linewidths with high damping, or large values of γ. Natural radiative lifetime broadening sets a lower limit on the linewidth. To account for both lifetime broadening and dephasing collision broadening, we replace γ in Eq. (4.16) with the quantity $(\gamma + 2/T_2)$.

Collisions can further broaden the atomic transition linewidth by increasing the rate at which energy leaves the atomic system. Collisions of this kind are analogous to inelastic collisions in classical mechanics: In a close interaction with other atoms, with the walls of a gas laser, or with a surrounding crystal lattice, the emitting atom gives up some of its stored energy, effectively increasing the energy damping constant. Such collisions shorten the lifetime of the transition, and the associated spectral broadening is consequently referred to by the general term *lifetime broadening*. Although the calculations of the size of the effects are difficult, the resultant damping constants from additional lifetime broadening mechanisms can be added to the radiative damping constant to obtain

$$\gamma_{total} = \gamma_{radiative} + \gamma_{other}. \tag{4.17}$$

The reciprocal of γ_{total} is the lifetime, τ, of the transition, which is the average length of time during which light energy is emitted by the atom. This quantity ranges up to several milliseconds in common laser materials.

Homogeneous versus Inhomogeneous Broadening

Broadening mechanisms can be classified into two general categories, homogeneous and inhomogeneous. A broadening mechanism is called *homogeneous* if every atom in the collection has the same transition center frequency and the same resonance lineshape. If, on the other hand, different atoms or groups of atoms within the collection exhibit different resonance frequencies or lineshapes for the same transition, the responsible broadening mechanism is referred to as *inhomogeneous*. In this second case, the medium responds to a larger range of frequencies than the linewidth of any single atom.

From these definitions, we see that collision broadening is a homogeneous broadening mechanism: Each atom in the gas has its spectrum spread in the same way by collisions (at least in a time average sense), and each atom has the same resonance frequency. Doppler broadening, on the other hand, is an example of inhomogeneous broadening, since individual groups of atoms in the collection have different apparent resonance frequencies depending upon their velocities relative to the observer. Local variations in temperature, pressure, and magnetic field applied to a material, as well as local variations due to crystal imperfections, also lead to inhomogeneous broadening of emission or absorption lineshapes. In some cases, both types of broadening are present in a collection of atoms and may contribute comparable amounts to the broadening.

Because of the various broadening mechanisms just discussed, we can no longer treat a group of radiating atoms as if all of them emitted at one particular frequency. Instead, we must consider a small spread of frequencies about some central value. The output of a laser might be expected to contain the same distribution of frequencies as the broadened transition of the atoms in the medium. In fact, it does not. The spectral character of the laser output is not the same as that for spontaneous emission in the same medium. Two effects contribute to this difference: the effects of the optical resonator, discussed in the next section, and the effect of the amplification process on the spectral irradiance. As light travels through an inverted medium, the spectral irradiance varies as

$$I_v(v,z) = I_v(v,0)e^{\beta(v)z}. \qquad (4.18)$$

Using Eq. (4.2) in the above expression, we see that $I_v(v,z)$ is related exponentially to the broadened atomic lineshape, $g(v)$. As a consequence, the function $I_v(v,z)$, which describes the amplified radiation, is stronger in the center and weaker in the wings, and therefore narrower in linewidth than the atomic lineshape. This effect, known as *spectral narrowing*, is one criterion used by researchers to determine if laser action is occurring in a medium. Thus, even without any attempts at mode selection or mode suppression (to be discussed in Section 4.2), laser light has a higher degree of monochromaticity than spontaneously emitted radiation from the same medium.

4.2 LASER MODES: OPTICAL RESONANCE

More dramatic than the spectral narrowing we have just discussed is the effect of the laser mirrors on the output of the laser. Using a Fabry-Perot interferometer like the one described in Section 2.6 to observe the output of a typical laboratory laser, one can show that this output consists of a number of discrete frequency components (very narrow spectral lines) and not the single broadened line that we have described. How do these discrete lines arise, and how are they distributed with respect to the laser transition lineshape? To answer these questions it is necessary to examine the effect of the mirrors on the light within the laser cavity.

Axial Modes of a Laser

The optical cavity of a laser is a resonator with extremely high Q (see box) and low losses. If these losses are smaller than the gain in the amplifying medium, threshold is achieved and lasing occurs. But the high-Q condition does not hold for all frequencies within the laser emission linewidth; only certain frequencies fulfill the resonance conditions, similar to the transmission conditions of the Fabry-Perot interferometer. Thus the laser output spectrum does not resemble the spontaneous emission lineshape, but rather consists of a series of narrower lines corresponding to the high-Q frequencies of the laser cavity.

To determine the conditions for high Q in a laser, we start with a plane wave of light propagating along a line normal to and between two parallel mirrors, just as we did with the Fabry-Perot interferometer discussed in Chapter 2. The round-trip distance for a wave undergoing reflection at the mirrors is $2L$, twice the distance between mirrors. The total phase change, $\Delta\varphi$, undergone by the wave in traveling a full round trip is equal to 2π times the distance traveled divided by the wavelength:

$$\Delta\varphi = 2\pi \frac{2L}{\lambda} = \frac{4\pi L}{\lambda}. \tag{4.19}$$

If the reflected wave is $180°$ out of phase with the original wave and of equal magnitude, then within the cavity there is no net field and therefore no net energy density to stimulate the atoms to emit, even if a suitable population inversion exists. The most useful way of viewing such a situation is to note that the wave has not replicated itself upon reflection. Only at such a frequency that the wave and its reflections are in phase ($\Delta\varphi = 2\pi q$, q an integer) does the wave replicate itself. With replication, the electric fields add in phase. The resultant energy density is sufficient to induce substantial stimulated emission at that frequency. From an alternative point of view, the mirrors form a resonant cavity in which light energy may be stored by multiple reflections between them. If the waves are replicated in the cavity, then the mirror cavity has a high Q. The condition for a self-repeating field (setting $\Delta\varphi = 2\pi q$ in Eq. 4.19) is that the length of the cavity

---

Q, THE QUALITY FACTOR

Fundamental to a discussion of any resonator is the concept of the Q, or *quality factor*, of the resonator, defined by

$$Q = \frac{2\pi \times \text{energy stored}}{\text{energy dissipated per cycle}}.$$

This definition of Q is a very general one and applies to circuits, mechanical systems, microwave cavities, and laser cavities. A typical oscillating electrical circuit, such as one containing a resistor, capacitor, and inductor, can have a Q of several hundred; a laser cavity can have a Q as high as 10^5 or 10^6. A high-Q cavity stores energy well, whereas a low-Q cavity does not. In addition, we note that a high Q is associated with a small relative linewidth, and a low Q with a larger relative linewidth. This relationship between Q and linewidth can be expressed rather simply as

$$Q = \frac{\text{resonant frequency}}{\text{linewidth}} = \frac{\nu}{\Delta\nu}.$$

be equal to an integral number of half-wavelengths, or $L = q(\lambda/2)$, q an integer. Only at those wavelengths is the cavity resonant. The integer q is in most cases quite large. For example, if the central wavelength is 500 nm (5×10^{-5} cm) and the mirror separation is 25 cm, q has a value of 10^6. Since q can be any integer, there are many possible wavelengths within the laser transition lineshape for which the field is self-replicating. We refer to each such self-replicating field pattern as a *longitudinal mode*, or *axial mode*, of the cavity. It is easier to refer to these axial modes by their frequency than by their wavelength. Using the condition for the self-replicating field stated above, we have

$$v = \frac{c}{\lambda} = \frac{c}{\left(\dfrac{2L}{q}\right)} = q\left(\frac{c}{2L}\right). \tag{4.20}$$

Each mode frequency can be labeled with its corresponding integer, q, with the result

$$v_q = q\left(\frac{c}{2L}\right). \tag{4.21}$$

It is at these frequencies that the laser cavity is resonant.

By subtracting the frequency of one cavity mode from that of its nearest neighbor, we find that the separation between mode frequencies is

$$\Delta v = v_{q+1} - v_q = (q+1)\frac{c}{2L} - q\frac{c}{2L}$$

or

$$\Delta v = \frac{c}{2L}. \tag{4.22}$$

The separation between longitudinal mode frequencies is seen to be the same as the free spectral range of a Fabry-Perot interferometer with plate separation L (see Eq. 2.32). Note that the separation between neighboring modes is dependent only on the separation between mirrors and is independent of q. If we use values from our example above, the separation between neighboring resonance frequencies for a typical laser (25 cm long) is calculated to be

$$\Delta v = \frac{3 \times 10^{10} \text{ cm/sec}}{2 \times 25 \text{ cm}} = 6 \times 10^8 \text{ sec}^{-1} = 600 \text{ MHz}. \tag{4.23}$$

Many laser transition lines are much broader than 600 MHz, and thus there can be many axial modes ($\ldots q-2, q-1, q, q+1, q+2, \ldots$) within the broadened linewidth. Since sustained laser action can occur only at those frequencies within the lasing transition for which the cavity is resonant, the output of a laser contains a number of discrete frequencies, separated by $c/(2L)$, as shown in Fig. 4.6. These frequencies are called the *axial mode frequencies* of the laser.

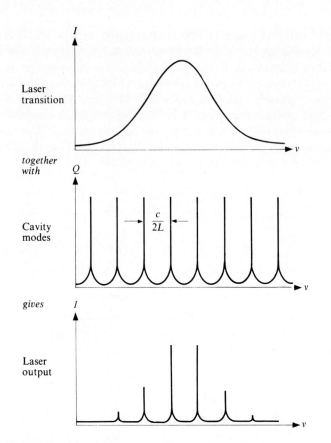

Fig. 4.6 The combination of the lasing transition lineshape with the resonant cavity modes gives the resulting output of a laser. Only when the Q of the cavity is high can lasing occur.

Longitudinal-Transverse Modes of a Laser

Thus far we have restricted our attention to resonance conditions for plane waves traveling along a line joining the centers of the mirrors, i.e., along the optic axis of the cavity. For any real laser cavity, the complete wave pattern consists of a superposition of a large number of waves, each wave traveling in a slightly different direction, only nominally along the optic axis. Under these circumstances, the resonance condition is more complicated. Once again, the basic requirement is that the electromagnetic field distribution in the cavity replicate itself upon round-trip reflection by the mirrors. Because of their three-dimensional nature, we refer to these self-replicating field patterns as the *longitudinal-transverse resonance modes* of the laser cavity.

It is no longer possible to characterize a cavity mode by a single number, q, since the transverse character of the mode must be taken into consideration as well as the axial character. Through an application of diffraction theory, it is possible to determine those field distributions that exhibit this self-replicating property in a laser cavity and to determine the frequencies of the corresponding modes. We simply quote the result; an extended analysis is found in Ref. 4.5. Recalling the g parameters for optical cavities defined in Eq. (3.28), the frequencies satisfying the full three-dimensional resonance condition can be expressed by the relationship

$$\nu_{mnq} = \left(q + (m + n + 1) \frac{\cos^{-1} \sqrt{g_1 g_2}}{\pi} \right) \frac{c}{2L} \tag{4.24}$$

where m, n, and q are integers. The sign of $g_1 g_2$ is positive if g_1 and g_2 both have the same sign; otherwise the negative sign is taken. This expression can be simplified for the common case of identical near-planar mirrors. Under these circumstances, $g_1 = g_2$ and $L/R \ll 1$, we obtain the simpler expression,

$$\nu_{mnq} \cong \left(q + \frac{(m + n + 1)}{\pi} \sqrt{\frac{2L}{R}} \right) \frac{c}{2L}. \tag{4.25}$$

In both Eqs. (4.24) and (4.25), the number referred to previously as the *axial mode number*, q, is once again associated with the axial character of the mode. On the other hand, m and n relate to the transverse character of the mode and are accordingly referred to as the *transverse mode numbers*. A mode of a laser is thus characterized by three mode numbers. In practice, the term "axial mode" is often used to differentiate between cavity modes with different values of q. In the same way, the term "transverse mode" is used to differentiate between modes of different m and n values. (The problem with the term "transverse mode" is that it can refer to many "axial modes" with the same m an n values.) The different modes are designated by the notation TEM$_{mnq}$, where TEM denotes Transverse Electro-Magnetic, the light waves consisting of electromagnetic fields that are transverse to the direction of propagation. As before, the value for q is quite large for practical laser dimensions. The precise value is not easily found and is generally of little interest. The values of m and n, by way of contrast, are usually quite small and are often determined by an inspection of the laser output. As a consequence, in the labeling of modes with specific values of m, n, and q, the q value is generally suppressed, with the resultant designation having the form TEM$_{mn}$. It is important to remember that although the mode designation does not contain q, each mode still retains a longitudinal character corresponding to some specific value of q. The important consideration is generally how many longitudinal modes (i.e., how many different values of q) are present in a laser output rather than their specific q-values, since it is the former quantity that determines the total spread in the laser output spectrum.

The mode structure of a laser—the number and nature of the modes that make up the laser output—is important in determining the potential uses of the laser. For example, certain transverse modes have large divergence and are undesirable if the laser output is to be tightly focused. However, the significance of laser modes goes far beyond such questions as divergence, because it is the self-replication of the electromagnetic fields within the laser cavity that results in the high degree of spatial and temporal coherence of the laser output.

It should be pointed out that our depiction of the growth of the electromagnetic fields as they propagate back and forth in the cavity is not altogether complete. An alternate picture is that of a self-consistent distribution of fields that fulfills the boundary conditions of the cavity and grows out of many other competing distributions that decay because the cavity is not resonant for them. In either case—the simple replicating wave picture or the self-consistent field distribution—the existence of modes is fundamental to the generation of light with the extraordinary properties described in Chapter 2.

Examining the Mode Structure of a Laser

The longitudinal mode structure of a laser output can be investigated with a Fabry-Perot interferometer. The experimental arrangement is shown in Fig. 4.7. You may recall from Chapter 2 that there are many frequencies, spaced $c/2d$ apart, that are transmitted by the interferometer when its plates are spaced a distance d apart. To block all extraneous light at frequencies other than in a narrow region about the laser line, we insert a narrowband filter between the laser and the interferometer. The spacing of the plates is then adjusted so that the free spectral range of the interferometer exceeds the linewidth of the laser. The transmitted light is focused by a lens onto a screen, where a pattern of concentric circles can be seen. A pinhole in the screen is positioned at the center of the pattern; light of one frequency passes the screen whereas other frequencies are blocked. As the spacing of the mirrors is changed, the frequency illuminating the pinhole changes. If d is changed in a continuous manner, the frequency passed by the pinhole is swept through a range of frequencies, which includes those of the laser line. A photomultiplier detector measures the amount of light transmitted at the different frequencies. An increase in signal occurs whenever the Fabry-Perot resonance scans through a frequency component of the laser output. If the output of the detector is plotted as a function of time, the observed result is a series of lines within the broadened transition lineshape, as sketched in Fig. 4.6. Each of these lines corresponds to a different axial mode, or q-value, of the laser. Knowing the interferometer scanning rate, one can verify that the frequency separation Δv between these modes is $c/2L$, where L is the distance between laser mirrors, in agreement with the analysis above.

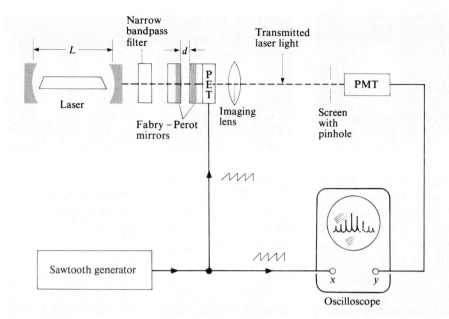

Fig. 4.7 Experimental arrangement for observing the longitudinal (or axial) mode characteristics of a laser. Light from the laser is filtered to remove all but the laser light. The separation between the Fabry-Perot interferometer mirrors is changed by applying a sawtooth voltage to a piezoelectric transducer (PET) attached to one of the mirrors. The change in separation d changes the transmission frequency of the interferometer. The transmitted light is focused onto a pinhole in a screen and detected by a photomultiplier tube (PMT). The output of the PMT is displayed on an oscilloscope. The oscilloscope trace displays the frequency spectrum of the laser.

Whereas sophisticated equipment is required to observe the longitudinal character or mode structure of the laser output, the transverse character can be easily identified by placing an inexpensive diverging lens in the beam and observing the pattern of the expanded beam on a screen. Typical patterns are illustrated in Fig. 4.8. These transverse mode patterns depend only on m and n, not on q. Notice that the patterns associated with each transverse mode are different and easily distinguished. The TEM_{10}^{*} mode pattern is a combination of the TEM_{01} and TEM_{10} patterns and may occur when there is a small lossy region or obstruction on the optic axis. It is given the rather descriptive name, the "doughnut mode." The pattern seen most often corresponds to the TEM_{00} mode, which has a Gaussian irradiance pattern (see Section 2.3) and the smallest beam divergence of any of the modes.

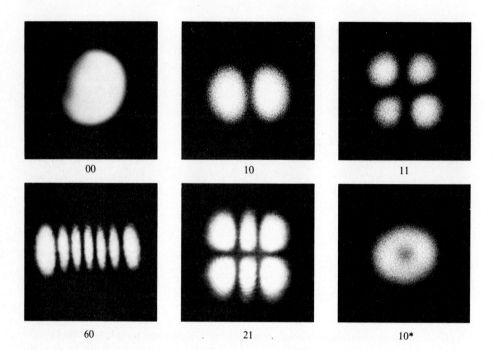

Fig. 4.8 Expanded beam patterns illustrating the different transverse characteristics of the laser modes. The numbers designate the corresponding values of m and n. (Photographs courtesy of Bell Telephone Laboratories and Spectra Physics, Inc.)

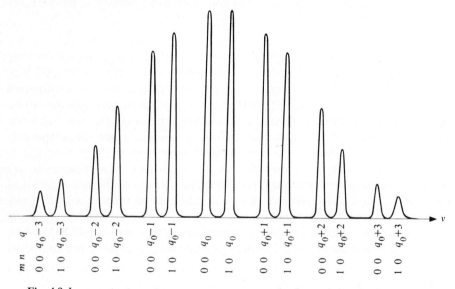

Fig. 4.9 Laser output spectrum, two transverse modes for each longitudinal mode.

It is possible for a laser to operate in more than one transverse mode, just as it is possible for a laser output to contain more than one axial mode frequency. The expanded beam patterns are then a combination of the contributing separate patterns. Under such circumstances, a Fabry-Perot interferometer scan would show a distribution of frequencies as in Fig. 4.9. Note that although two modes may have the same q-value (i.e., be identified with the same longitudinal mode), the frequencies can still be different, in agreement with Eq. (4.25). In the figure, each line is labeled to show that there are sixteen separate modes present. Each longitudinal mode has associated with it two different transverse modes, TEM_{00} and TEM_{10}.

How many lines can there be beneath the lasing transition? The number of longitudinal modes is determined by the linewidth of the transition line and by the length of the laser: The longer the laser, the smaller the separation between modes of different q-values, and thus the greater the number of modes present within the laser transition linewidth. The number of transverse modes will depend on mirror shape, size, and other aspects of laser construction. When there are a number of modes in the laser output, we refer to the laser as operating *multimode*.

By placing appropriate devices in the laser cavity, one can often modify the laser output. For example, by introducing selective losses for certain transverse modes, one can force a multimode laser to lase with a specific transverse mode pattern. In some lasers, an adjustable iris diaphragm is located on the optic axis of the cavity to suppress all transverse modes except those with TEM_{00} character. This is because other modes, which have some off-axis component, experience more loss with the diaphragm closed down than do the TEM_{00} modes.

4.3 PUMP RATE, GAIN OSCILLATIONS, AND POWER OUTPUT

Having developed some understanding of the modes of a laser, we can now examine more closely the connection between the degree of population inversion in the laser cavity, the laser gain, and the output of the laser. Our examination will be based on a study of a four-level laser system. In spite of certain simplifying assumptions in the analysis, the results are applicable to many laser systems.

In Fig. 4.10, we show again the energy state populations in a four-level system. The numbers of atoms undergoing transitions between the various states each second (Section 3.1) are indicated. The predominant transitions are shown in the diagram next to the four-level population distribution. Secondary transitions (e.g., $3 \rightarrow 1$), assumed negligible or forbidden, are shown in the right-hand diagram. The laser is assumed to possess only a single high-Q cavity mode to prevent possible competition between cavity modes from increasing the difficulty of the analysis. The *pump rate*, R, is the number of atoms per second undergoing transitions from the ground state to the highest energy state. In an electrical discharge-pumped laser, R increases with increasing supply current; in a flashlamp-pumped laser, R increases with lamp intensity.

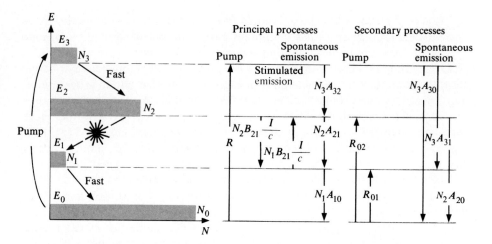

Fig. 4.10 Population of energy levels in a four-level system. The figure in the center indicates the principal processes considered in the simple analysis of the system, whose population distribution is shown at the left. The figure on the right shows other processes that serve to reduce the overall gain of the lasing system. These secondary processes have not been included in the analysis given in the text.

Steady-State Case

It is possible to express the rate of change in population of each of the four states in terms of the number of transitions to and from each state, as labeled in Fig. 4.10. The differential equations so obtained are known as *rate equations*. These equations can be reduced to simple algebraic equations if steady-state conditions are assumed—i.e., assume that the system is being pumped at a constant rate. This is equivalent to assuming that the time derivatives of the populations of the levels are all equal to zero. The resultant algebraic equations can be manipulated to yield an expression relating the population inversion, $N_2 - N_1$, to the pump rate, R, and to the beam spectral irradiance I within the cavity:

$$N_2 - N_1 = \frac{R\left(1 - \dfrac{A_{21}}{A_{10}}\right)}{A_{21} + B_{21}\left(\dfrac{I}{c}\right)} \tag{4.26}$$

where A_{21}, A_{10}, and B_{21} are the appropriate Einstein coefficients for the transitions among energy levels 0, 1, and 2. We see that unless $A_{21} < A_{10}$, the quantity in the numerator is negative, and no population inversion can take place. Recall that $A_{ij} = 1/\tau_{ij}$, i.e., the spontaneous transition rate between states i and j is the reciprocal of the spontaneous emission lifetime for the $i \to j$ transition. Thus the condition $A_{21} < A_{10}$ is equivalent to the condition $\tau_{10} < \tau_{21}$, i.e., that the upper state of the laser transition has a longer spontaneous emission lifetime than the lower state. In most laser systems it is much longer, and the quantity $1 - (A_{21}/A_{10})$ is near unity.

Below threshold, the stimulated emission from the upper level of the laser transition (energy E_2) is small, because the beam irradiance I at the laser frequency is itself very small. The power that is pumped in is squandered on all the modes in the laser cavity, both high- and low-Q, and on the spontaneous emission at the laser frequency. Thus, below threshold,

$$N_2 - N_1 \approx \frac{R\left(1 - \dfrac{A_{21}}{A_{10}}\right)}{A_{21}} \tag{4.27}$$

that is, there is a linear increase in population inversion with pumping rate, but not sufficient inversion to maintain amplification.

At threshold, I is still small, and the threshold population inversion is

$$\Delta N_{\text{th}} = (N_2 - N_1)_{\text{th}} = \frac{R_{\text{th}}\left(1 - \dfrac{A_{21}}{A_{10}}\right)}{A_{21}}. \tag{4.28}$$

And in steady state, the population inversion never gets any larger than this! Remember that the small signal gain β is linearly dependent on the population inversion (Eq. 3.20). And as we pointed out in Section 3.3, the gain of the laser must equal the losses; otherwise the irradiance within the cavity would continue to increase with time, a violation of our steady-state assumption. This phenomenon, known as *gain saturation*, was discussed in Chapter 3. The requirement stated there that the small signal gain be equal to the threshold gain can be restated as $(N_2 - N_1)_{\text{ss}} = \Delta N_{\text{th}}$ for a laser operating steady state above threshold. The population inversion then enters Eq. (4.28) as a constant, giving

$$\Delta N_{\text{th}} = \frac{R\left(1 - \dfrac{A_{21}}{A_{10}}\right)}{A_{21} + B_{21}\dfrac{I}{c}} \qquad \text{whenever} \qquad R \geqslant R_{\text{th}} = \frac{\Delta N_{\text{th}} A_{21}}{\left(1 - \dfrac{A_{21}}{A_{10}}\right)}. \tag{4.29}$$

The only variables are the pump rate, R, and the beam irradiance, I, of the high-Q laser cavity mode.

The beam irradiance can be expressed as a function of the pump rate R:

$$I_{\text{cavity}} = R \left[\frac{\left(1 - \dfrac{A_{21}}{A_{10}} \right)}{A_{21}\, \Delta N_{\text{th}}} - 1 \right] \frac{4\pi c}{B_{21}} A_{21} \tag{4.30}$$

The laser irradiance increases linearly with pump rate once threshold is reached. Plots of ΔN and I_{cavity} as a function of pump rate R (Fig. 4.11) indicate the relationship between these quantities in this highly simplified situation. The additional power above threshold has been channeled into the single (or a few) high-Q cavity mode that dominates the laser output. The spontaneous emission background does not disappear above threshold, but compared to the laser output it is extremely weak, since it is emitted in all directions. Also, the spontaneous emission is spread in frequency over the entire emission lineshape, whereas the laser radiation is concentrated at the mode frequencies.

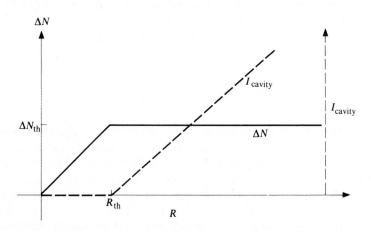

Fig. 4.11 Population inversion and laser power as a function of pump rate.

Time Dependent Case-Spiking

Our derivations of the above equations depend upon the assumption that the small-signal gain, β, is a time-independent quantity. Such an assumption may not be valid, however, with strong pumping in certain lasing systems. To obtain a clearer picture of the time dependence of the intracavity amplification, we must relate the growth of irradiance within the laser cavity not only to β, but also to the dynamic changes in the atomic-state populations. Let us assume we have a propagating wave in an inverted medium before gain saturation sets in. The irradiance grows exponentially with time as the wave travels through the medium:

$$I = I_0 e^{(\beta - \alpha_t)ct} \tag{4.31}$$

in accord with Eq. (3.24). By differentiating the above equation and substituting for β from Eq. (3.20), the time dependence of I can be shown to be

$$\frac{dI}{dt} = k_1 \, \Delta N I - k_2 I \tag{4.32}$$

where k_1 and k_2 are constants of the active medium. The population difference ΔN is also time dependent and obeys a relation

$$\frac{d(\Delta N)}{dt} = k_0 - k_3 \, \Delta N - k_4 \, \Delta N I \tag{4.33}$$

where k_0 is proportional to the pump rate and k_3 and k_4 are constants of the active medium. These two equations (4.32 and 4.33) form a pair of coupled linear differential equations.

We would like to solve these equations for ΔN as an explicit function of time, because if we could do this, we could obtain an expression for $\beta(t)$, by substituting back into Eq. (3.20). The equations cannot be solved analytically as they are written. With certain approximations, however, analytic expressions for I and ΔN can be obtained. These expressions take the form of damped oscillations, which, for constant pump rate R, may ultimately settle down to steady-state values. The small-signal gain, β, being related as it is to ΔN, exhibits similar behavior. One can readily see how oscillations in β might occur. When a population inversion has been obtained and stimulated emission begins, the intracavity irradiance grows exponentially, provided ΔN stays constant. But the growth in irradiance comes about through the depopulation of the upper level of the laser transition, and this depopulation causes ΔN to decrease. We now recognize that the small-signal gain is a variable and that it decreases as ΔN decreases. Once β is reduced, the irradiance will decrease, ΔN will again increase, and oscillation can occur.

The precise behavior of the gain depends upon many factors, and varies drastically among lasers of different type, construction, and operation. The laser output power may oscillate or it may come to its steady-state value without oscillating. In many high-power, pulsed lasers, the depopulation of the excited state may be so large that β falls below the threshold level, and lasing stops until the population inversion can be restored by pumping, at which point the whole process begins again. The output of such lasers consists of a series of "spikes," as shown in Fig. 4.12. The spikes for a ruby laser are about one microsecond in duration with about one microsecond separation between spikes. If the pumping lasts several milliseconds, thousands of spikes occur in the duration of the pump pulse.

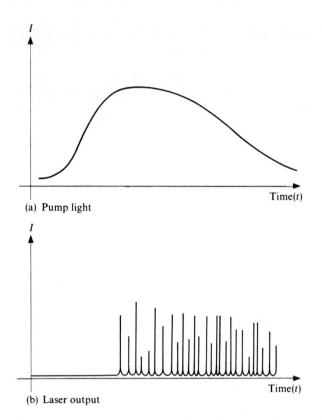

(a) Pump light

(b) Laser output

Fig. 4.12 Illustration of the time development of spiking after a laser is pumped with a large amount of light.

4.4 GAIN SATURATION IN HOMOGENEOUSLY AND INHOMOGENEOUSLY BROADENED LASERS

In many steady-state, or continuous-wave, lasers, there are initial oscillations in the small-signal gain. Eventually, however, β settles down to its threshold value, β_{th}, because of gain saturation, as discussed in Chapter 3. The manner in which gain saturation affects the laser output depends greatly upon the particular kind of broadening mechanisms present.

Recall that for lasers that are homogeneously broadened, the emission from each atom in the active medium is broadened identically, and thus all atoms have the same response to stimulation at any particular frequency. The effect of gain saturation on such a homogeneously broadened system is illustrated by example in Fig. 4.13. At the start of lasing, there are three cavity-mode frequencies for which the gain is above threshold. As laser oscillations build up at these fre-

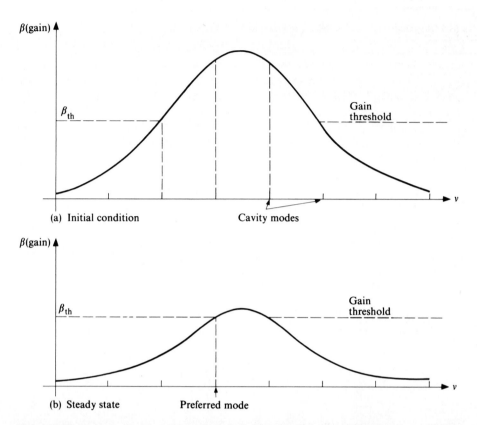

Fig. 4.13 Gain saturation in a homogeneously broadened laser. (a) At the start of lasing, three cavity modes have gain above the threshold. (b) When the steady state is reached, the gain saturates (reaches the threshold value) at that mode frequency with the highest gain.

quencies, the gain is reduced by saturation. The modes compete with one another for the inverted population, and as saturation increases, only one mode is left with a gain above threshold. As Fig. 4.13(b) indicates, the final steady-state saturation occurs when the gain of the preferred frequency is reduced to the threshold level. Thus, homogeneously broadened lasers should automatically operate in a single mode once a steady state is reached. This does not always occur, however, because it is possible for different mode frequencies to use different regions of the lasing medium to lase. For a particular longitudinal-transverse mode, there are nodal regions where little stimulated emission occurs, and thus the gain in these regions remains high. These regions may support laser action at another resonant cavity frequency. In the case of the ruby laser, which is an example of a homogeneously broadened system, special precautions must be taken to actually achieve single-mode operation.

Inhomogeneously Broadened Lasers—Hole Burning

Gain saturation of inhomogeneously broadened lasers is quite different from that for homogeneously broadened lasers. In the case of inhomogeneous broadening, the response of atoms emitting at one frequency of the lasing transition line or gain curve is independent of the response of atoms resonant at a different frequency. It is possible, therefore, for steady-state oscillations to occur simultaneously at several frequencies within the gain curve, as shown in Fig. 4.14. At those frequencies where the cavity is not resonant, the gain remains high, because the population inversion is at a maximum. At cavity resonance frequencies, however, the gain decreases to the threshold value, as it does in the homogeneously broadened laser. The cavity modes put "holes" in the gain curve and the process is consequently referred to as *hole burning*. If more cavity modes are introduced by multimoding the laser, the number of atoms that can participate in the laser oscillation is increased and output power increases. (The increased output is purchased at the expense of greater beam divergence and a slightly broader linewidth.) Should the distribution of cavity modes be sufficiently dense, it is possible to "burn off" the top of the entire gain curve.

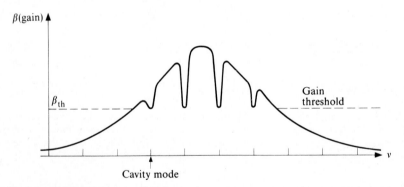

Fig. 4.14 "Hole burning." In an inhomogeneously broadened laser, different groups of atoms respond to the stimulating radiation fields at the different cavity-mode frequencies. The gain is saturated for each group of atoms independently, creating "holes" in the gain curve.

In this chapter, we have described the output of pulsed and continuous-wave lasers in their most basic configuration: a device with a population inversion of the energy states in an active medium contained within a mirror cavity. Not all the characteristics of the laser output are desirable for certain applications. We will now look at methods of modifying the output of the laser based upon the principles of Chapter 3 and the descriptions from this chapter.

PROBLEMS

4.1   On the basis of the definitions presented, state whether natural lifetime broadening is an example of homogeneous or inhomogeneous broadening. Why?

4.2   The presence of a strong magnetic field will sometimes split atomic resonance spectral lines into two components, the amount of the splitting being proportional to the magnetic field strength (Zeeman effect). In the figure below, a collection of such atoms is placed between the wedge-shaped poles of a powerful magnet and the absorption spectrum observed. Will the resonance lineshape exhibit homogeneous or inhomogeneous broadening? Explain.

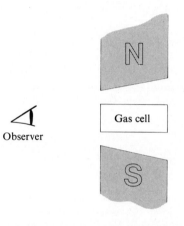

4.3   At low pressures, the absorption lineshape associated with a gas appears to be Doppler-broadened, whereas at high pressures, the same gas at the same temperature exhibits predominantly collision broadening (sometimes referred to appropriately as pressure broadening). Explain in words why this might be the case.

4.4   Compare the Doppler-broadened linewidth of a CO_2 laser transition $\lambda = 10.6$ μm with that of the neon line $\lambda = 0.633$ μm discussed in the text. Assume that the CO_2 laser gas discharge temperature is also about 400 K.

4.5   Show by substitution that

$$x(t) = x(0)e^{(-1/2)\gamma t} \cos 2\pi v_a t \quad \text{where} \quad v_a = \left[v_{ij}^2 - \left(\frac{\gamma}{4\pi} \right)^2 \right]^{1/2}$$

is the exact solution to Eq. (4.13) and that the expression in Eq. (4.14) is thus approximately correct for $v_{ij} \gg \gamma$. What is the order of magnitude of the fractional error, $(v_{ij} - v_a)/v_{ij}$, in the approximate resonance frequency calculation, assuming $v_{ij} = 10^{14}$, $\gamma = 10^6$?

4.6   Assuming that $v_{ij} \gg \gamma$, show that the time average of $x^2(t)$ (where $x(t)$ is given by Eq. 4.14), and hence the oscillator energy, decays exponentially with a rate γ.

4.7   Show that γ is equal to 2π times the half-power linewidth of the Lorentzian spectral irradiance distribution defined in Eq. (4.16).

4.8   Assume that a particular optical transition of an atom is characterized by a resonance frequency of 5×10^{13} Hz and a broadened lifetime, τ, of 1 μsec (where $\tau = 1/\gamma$). How many oscillatory periods occur for the atom before the stored energy decays to the $1/e$ value?

4.9   What is a piezoelectric transducer? How does it work and what is it made of?

4.10   On the basis of the patterns shown in Fig. 4.11, sketch patterns corresponding to the TEM_{22}, TEM_{41}, and TEM_{32} transverse modes.

4.11   Assuming the center wavelength of the helium-neon laser is exactly 632.8 nm, what is the corresponding frequency? If the cavity mirror separation is exactly 50 cm, what is the value of q for the laser mode nearest the line center? If the width of the gain curve of the helium-neon line were 1 GHz, what is the number of longitudinal cavity modes inside the line-width for the cavity above?

4.12   What is the frequency difference between a TEM_{00} mode and a TEM_{01} mode with the same q value, assuming $L = 40$ cm and $R_1 = R_2 = 100$ cm. If this is a helium-neon laser, what fraction of the frequency does this difference represent? Which frequency is higher?

4.13   Sketch curves of the gain (β), the output irradiance (I), and the population difference (ΔN) as a function of time to illustrate spiking.

4.14   The rate equation for the upper level of the laser transition of the four-level system shown in Fig. 4.10 is given by

$$\frac{dN_2}{dt} = N_1 B_{21} \frac{I}{c} - N_2 B_{21} \frac{I}{c} - N_2 A_{21}$$

assuming only principal processes are considered. Write down the equations for the population rate of the other three levels of the system. Under steady-state conditions (time derivatives are zero), use the resulting algebraic equations to obtain Eq. (4.26).

REFERENCES

4.1 B. A. Lengyel (1971), *Lasers* (2nd edition). New York: Wiley.

4.2 A. Maitland and M. H. Dunn (1970), *Laser Physics*. New York: American Elsevier.

4.3 R. H. Pantell and H. E. Putoff (1969), *Fundamentals of Quantum Electronics*. New York: Wiley. Excellent discussion of the rate equations that requires no quantum-mechanics background.

4.4 Editors of *Scientific American* (1969), *Lasers and Light*. San Francisco: W. H. Freeman. Reprints of *Scientific American* articles with introductions to major sections by A. L. Schawlow. Nonmathematical readings. The laser-related articles occupy the final third of the book. Some of the sections are dated.

4.5 A. E. Siegman (1971), *Introduction to Lasers and Masers*. New York: McGraw-Hill.

4.6 O. Svelto (1976), *Principles of Lasers*. New York: Plenum.

4.7 A. Yariv (1976), *Introduction to Optical Electronics* (2nd edition). New York: Holt, Rinehart and Winston.

4.8 A. Yariv (1975), *Quantum Electronics* (2nd edition). New York: Wiley.

Modifying
the Laser Output

From the discussions of the last chapter, it is now evident that the output of the laser is not as simple as it was portrayed in Chapter 3. The output spectrum may consist of a large number of discrete frequency components. In addition, the time evolution of the laser action may produce an output consisting of a series of irregular spikes. For many laser applications, such spectral and temporal characteristics are quite acceptable. It is often desirable, however, through modification of the basic laser system, to obtain outputs with spectral and temporal characteristics better suited to certain applications. For example, in the previous chapter we mentioned that higher-order transverse modes in a laser could be suppressed by inserting a small circular aperture or adjustable diaphragm in the cavity. Thus by increasing the losses for off-axis modes, the laser can be modified to operate only in the TEM_{00} mode. Since the TEM_{00} mode has lower beam divergence than other modes, such a modification might be required, for example, if one were to use the laser for an alignment application. In this chapter, we investigate the theory behind various modifications and look at some specific examples.

5.1 SELECTION OF LASER EMISSION LINES: INTRACAVITY ELEMENTS

Many lasers can lase in several emission lines simultaneously. Here we are speaking not of the multiple-mode structure beneath a single broadened emission line, but of several of these lines spaced across the spectrum. While this multiple-line emission may provide high-power output, there are cases where a higher degree of monochromaticity is desired. In these cases, a wavelength-dependent element, which disperses or absorbs the light according to its colors, is introduced into the cavity. This element can be a prism, a grating, or a filter.

The action of these elements can be demonstrated by the example of an intra-cavity prism. The arrangement is shown in Fig. 5.1. We assume that three wavelengths, λ_1, λ_2, and λ_3, are undergoing amplification by stimulated emission, with $\lambda_1 < \lambda_2 < \lambda_3$. The shorter wavelengths suffer greater refraction by the prism than the longer wavelengths. One beam, at λ_2, is directed normally to the end mirror and then retraces its path back into the laser tube. Rays at other wavelengths do not retrace their paths exactly and, consequently, experience additional loss. Only for light in a small wavelength range about λ_2 are the losses smaller than the available gain. By using a properly cut prism, the angles of incidence can be set at the Brewster angles, reducing the losses due to reflection at the prism surface.

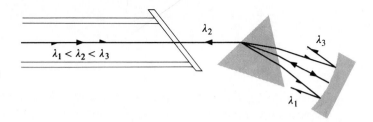

Fig. 5.1 Intracavity prism for wavelength selection. The prism disperses the light so that only one ray, at λ_2, is reflected back into the active medium, and lasing occurs only at the selected wavelength.

5.2 SINGLE-MODE OPERATION

When operated in a single longitudinal, single transverse mode, a laser is the closest approximation to a monochromatic light source known. The linewidth of a single laser mode is far smaller than the broadened transition linewidth; in some cases smaller than the linewidth due to the natural lifetime of the excited state. Since an inhomogeneously broadened laser can support many longitudinal-transverse modes simultaneously, single-mode output can be achieved only by assuring that one mode has a gain higher than all the others. There are several methods for obtaining single-mode output, two of which we discuss here. To begin with, let us ensure that the cavity supports only a single transverse mode, the TEM_{00} mode, by placing apertures in the laser cavity as we discussed earlier. Once this single transverse mode has been obtained, the problem is to eliminate all but one of the axial modes. One way to achieve single axial-mode operation is to design the cavity so that only one axial mode is possible within the laser transition linewidth. If the mode corresponding to q_0 is within the transition linewidth and those corresponding to $q_0 + 1$ and $q_0 - 1$ are outside it, as shown in Fig. 5.2, then only the TEM_{00q_0} mode will lase. For this to occur, the distance between cavity modes, $c/2L$, must be somewhat greater than the broadened linewidth. Since we

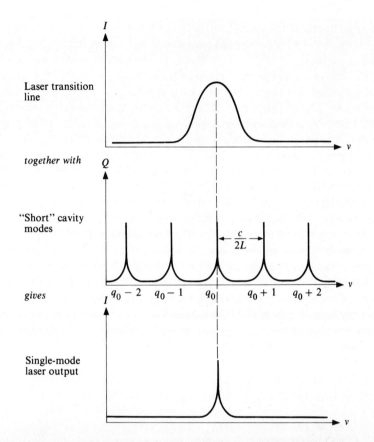

Fig. 5.2 The "short-cavity" method of single-moding a laser. If the length of the laser cavity is reduced to a length that yields a cavity-mode separation somewhat greater than the linewidth, only one cavity mode can lase.

cannot control the width of the lasing line, we must construct the cavity in such a manner that $c/2L$ is larger than the laser linewidth. If we make the distance between mirrors sufficiently small so that only one mode is supported, we produce a single-mode laser.* The drawback of this method is that the active length of the laser is also small, severely limiting the power output.

---

* To maintain the single-frequency output at a constant level, we must also stabilize the dimensions of the cavity through rigid cavity construction and compensating systems. If this is not done, L will change and the power and frequency of the laser output will change. In passing, we note that it is possible to stabilize the operating frequency of a laser to about 7 MHz, or about one part in ten million!

Another method for obtaining increased single axial-mode output from a TEM$_{00}$ laser is to introduce large losses for all but one of the modes. This can be done by introducing a small fixed-spacing Fabry-Perot cavity within the laser cavity, as illustrated in Fig. 5.3. The additional cavity consists of a special piece of glass, called an *etalon*, that has two faces ground and polished to a high degree of parallelism. The etalon cavity differs from the laser cavity in two important respects. First, the etalon surfaces are either uncoated or lightly coated, and their reflectivity is thus quite low. Because of this low reflectivity, the etalon cavity resonances are broader than the laser cavity resonances. Second, the etalon cavity is much shorter than the laser cavity, and the separations between etalon resonant frequencies are therefore much larger than those between the laser resonant frequencies. The etalon, in effect, makes slats in the picket fence unequal: Some cavity resonances have a higher loss than others. The laser tends to lase in that single mode with the smallest loss. This single-mode selection is illustrated in Fig. 5.4. As much as 75 percent of the power distributed over all the axial modes present before the etalon is inserted can appear in this single mode in a typical laser. If the etalon is tilted with respect to the optic axis, the frequency of the etalon resonance shifts within the lasing transition linewidth. A different laser cavity mode becomes the high-Q mode, and lasing occurs at the corresponding new frequency. It is thus possible to "tune" the frequency of the laser over a narrow frequency band by merely tilting the etalon.

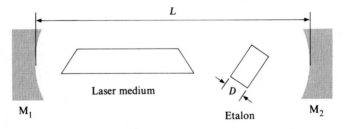

Fig. 5.3 The "etalon" method of single-moding a laser. Introduction of a piece of glass with parallel faces (the etalon) into the cavity renders the Q of the cavity modes unequal. Only the highest Q mode lases.

There are more complicated methods of obtaining single-mode output for inhomogeneously broadened laser transitions. Those we have discussed give some idea, however, of how a single laser mode is obtained and emphasize the importance of understanding the concept of laser modes.

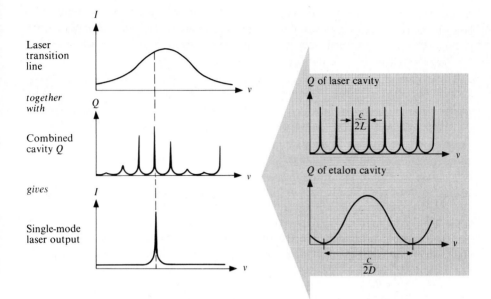

Fig. 5.4 The "etalon" method of single-moding a laser. The box on the right illustrates the Q versus frequency for the laser cavity and for the etalon cavity. Their combined Q versus frequency is shown in the middle curve. The combination of the lasing transition and a single high-Q mode results in a single-mode laser output.

The Lamb Dip and Laser Frequency Stabilization

In most gas lasers, the principal broadening mechanism is Doppler broadening, where the frequency of light emitted or absorbed by an atom is dependent upon the velocity of the atom, as described in Section 4.1. In single-mode lasers that have Doppler-broadened emission lineshapes, an interesting effect occurs. When the laser cavity frequency of the single mode is tuned to the peak of the gain curve, there is a drop in the output power compared to that obtained when the cavity is tuned off the peak of the curve. This effect, called the *Lamb dip*, was predicted on purely theoretical grounds by Willis E. Lamb and later verified experimentally. Its explanation requires an understanding of both Doppler broadening and gain saturation.

Let us suppose that we have a single-mode laser with cavity-mode frequency v_c that is greater than the natural emission frequency of the atoms, v_0. The light within the cavity, being reflected back and forth by the laser mirrors, consists of two sets of waves, one set traveling in the positive z-direction, the other traveling in the negative z-direction. Both of these waves are at the cavity frequency v_c.

Atoms having no velocity component along the optic axis are not stimulated to emit, because the laser frequency is not at the atomic resonance frequency. There is, however, a group of atoms with a positive z-axis velocity component that satisfies the Doppler relationship

$$v_0 = v_c \left(1 + \frac{|v_z|}{c} \right)^{-1}, \qquad v_c > v_0 \tag{5.1}$$

For this group of atoms, the apparent frequency of those waves traveling in the same direction is v_0 and the atoms are stimulated to emit. Thus for a very small range of velocities about v_z, the population inversion is reduced by the stimulated emission and gain saturation occurs. The above argument is also true for a second group of atoms and for waves moving in the negative z-direction. There are, then, two groups of atoms whose stimulated emission can contribute to the laser line intensity.

In Fig. 5.5(a), we have plotted the density of atoms in the excited state as a function of the z-component of the velocity. The stimulated emission produces a

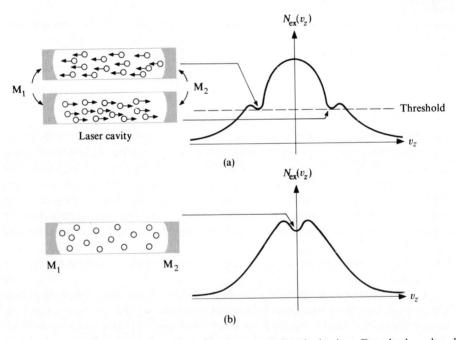

Fig. 5.5 Distribution of excited-state atoms versus z-axis velocity in a Doppler-broadened laser emission line. (a) The cavity frequency is not equal to the atomic resonant frequency. (b) The cavity frequency equals the atomic resonant frequency.

saturation in the *excited-state* velocity distribution, in a manner analogous to hole burning in the gain curve. The important point is that two groups of atoms with equal and opposite velocities are the only contributors to the lasing process. If the frequency of the cavity mode is changed (generally all that is needed is to vary the temperature of the laser cavity slightly), the cavity mode can be tuned to the peak frequency of the laser line, $v_c = v_0$. Under these circumstances, only a *single* group of atoms can contribute—those with no z-component of velocity (i.e., those atoms standing still or moving at right angles to the optic axis). These conditions are illustrated in Fig. 5.5(b). When this happens, the laser output power drops, since the available inverted-state population is smaller than before. If we plot the laser output as a function of v_c, as in Fig. 5.6, there is a dip (the Lamb dip) at $v_c = v_0$. Since any slight deviation from the center of the laser line results in an increase in power, the effect can be used in a feedback system to maintain the frequency of the laser at the line center by minimizing the output. This technique is used commercially, for example, to frequency stabilize helium-neon lasers. Because of the resulting long coherence length (tens of meters), these lasers are useful in applications like long path-difference interferometry and optical communications.

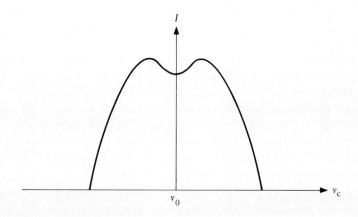

Fig. 5.6 Output of a single-mode laser as a function of cavity frequency. The minimum at the center of the curve is the Lamb dip.

The Lamb dip serves to bring together a number of important concepts we have discussed so far. The gain curve arises initially from the Doppler broadening of the laser transition; the frequency of the laser is determined by the single cavity mode; and the saturation effects on the velocity distribution produce the central dip in the irradiance as the cavity is tuned through the center of the gain curve.

5.3 Q-SWITCHING

In the first two sections of this chapter, we have emphasized the frequency dependence rather than the time dependence of the laser output. For many applications, however, the frequency characteristics of laser light are much less important than its time dependence. In this section, we look at some techniques for obtaining high-power pulses from a laser.

Single laser pulses of high peak power can be achieved by introducing into the laser cavity an irradiance-dependent or time-varying loss. The effects of such changing losses can be interpreted in terms of the gain and spiking discussions of Sections 3.3 and 4.3. If there is initially a high loss in the laser cavity, the gain from the inverted population can reach a very high value without laser oscillation beginning. Laser oscillation is prevented by the loss, while energy is pumped to the excited state of the active medium. When a large population inversion is reached, the cavity losses are deliberately and suddenly reduced. The threshold gain decreases immediately; the actual gain, however, remains high because of the large excited-state population. Because of the large difference between actual gain and threshold, the laser radiation in the cavity quickly grows, and all available energy is emitted in a single, large pulse. This pulse rapidly depopulates the excited state to such an extent that the gain is reduced below its threshold value, and lasing action ceases. The sudden altering of the losses of a laser cavity is known as Q-switching, because of the close relationship between resonator losses and resonator Q, as discussed earlier.

Q-switching has a dramatic effect on the output of solid state lasers. In the ordinary pulsed mode, the output of a ruby laser consists of many random spikes of approximately one microsecond duration with an interspike separation of about one microsecond. Peak powers obtainable in the pulsed mode are typically on the order of kilowatts; the length of the train of spikes is determined principally by the duration of the flashlamp excitation source. When the same laser is Q-switched, however, the result is one single spike of great peak power, typically in the megawatt range, with a duration of less than a hundred nanoseconds.

Rotating Reflector Method

The method first developed for Q-switching employs a rotating mirror or prism in place of one of the fixed mirrors of the laser cavity, as shown in Fig. 5.7. The other cavity mirror is a laser reflector capable of withstanding high power. The movable mirror revolves with high angular velocity. A low-power laser or a small light source is arranged such that its beam is reflected onto a detector just prior to alignment of the rotating mirror with the other cavity mirror. The detector emits a pulse that causes the flashlamp to fire, thereby exciting the laser. Since the mirror is not yet aligned when this occurs, the cavity losses are too large to sustain lasing action, and the population difference becomes considerably greater than would be the case if the mirror were aligned. After a large population difference and, therefore, high small-signal gain have been attained, the mirror moves into alignment, drastically reducing the losses, and the cavity Q switches to a very high value. The resulting high Q-high gain condition allows a rapid buildup of electromagnetic energy, and a Q-switched pulse results. The time evolution of this process is illustrated in Fig. 5.8. Although this figure specifically refers to the rotating reflector method of Q-switching, a thorough understanding of the relations between the various graphs extends to the other methods. Note that the cavity Q and the round-trip gain attain their maximum value when the mirror is aligned ($\theta = 0°$). At this point, the Q-switched pulse is emitted, with the resultant reduction of gain.

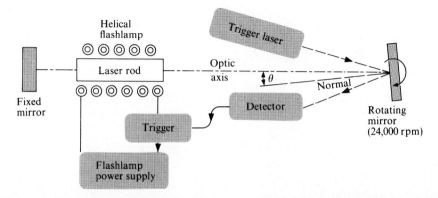

Fig. 5.7 Rotating reflector method of Q-switching. A small laser is positioned such that its light is reflected from the rotating reflector onto a detector, triggering the flashlamp just prior to alignment of the reflector.

It should be pointed out that the repetition rate of the laser firing is determined by electronic control circuits on the flashlamp and not by the mirror rotation speed. If the laser were fired on every revolution of the mirror, the repetition rate would be on the order of several hundred times per second. For large ruby lasers, this rate is prohibitive due to heating of the laser rod. The rotating reflector method of Q-switching has the advantages of extreme ruggedness and wavelength independence. In addition, some pulse-length control can be exercised by variation of the mirror rotation rate. However, this method is not the most efficient method for producing Q-switched laser pulses. Even though the mirror is spinning rapidly, its rate of rotation is still not high compared to the rate of decay of electromagnetic energy in the cavity. Basically, it is a "slow switch." This results in a less efficient production of Q-switched pulses, with lower peak power.

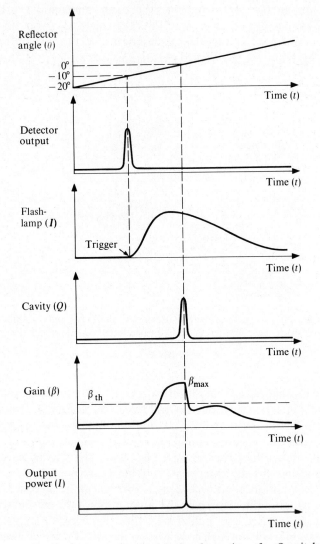

Fig. 5.8 Time evolution of laser parameters during formation of a Q-switched pulse using the rotating reflector method.

Electrooptic and Acoustooptic Shutters

If we consider the rotating reflector as both a laser mirror and a fast shutter combined in one element, other methods of *Q*-switching are easily understood. When the shutter is closed (corresponding to misalignment of the mirror), the laser cavity has a low-*Q*. When the shutter is opened (corresponding to alignment of the mirror), the *Q* of the laser cavity is switched rapidly to a high value. The essential feature of *Q*-switching is that the shutter is switched from closed to open when the gain of the active medium is near its maximum value.

Electrooptic modulators (see box) may be used as high-speed shutters to produce efficient *Q*-switching. Generally, there are two arrangements for producing

---

OPTICAL MODULATORS AND SHUTTERS

We noted in Chapter 2 that in birefringent crystals, light polarized along the "fast" axis of the crystal propagates with a higher velocity than light polarized along the "slow" axis. This property allows such crystals to be used to modify the state of polarization of a light beam — e.g., from linear to circular or from horizontal linear to vertical linear. *Electrooptic crystals* have the additional property that their birefringence (the difference in refractive index for "slow" and "fast" axis polarizations) can be varied in a controlled manner by the application of a voltage across the crystal. This characteristic makes them useful in electrically controlled optical shutters and modulators. Depending on the applied voltage, the electrooptic crystal serves as a quarter-waveplate (linear to circular polarization converter), a half-waveplate (rotates linear polarization through some angle), or a full-waveplate (no change). Combined with polarizers of the proper kind and orientation, the crystal can then control the amount of light transmitted in an optical system.

The *acoustooptic modulator* is based on the diffraction of light by sound waves in a medium such as glass or water. As shown in the figure, the sound wave is launched across the beam path by applying a sinusoidal voltage of high frequency to a piezoelectric transducer (PET) attached to the side of the medium. The sound waves create spatially periodic changes in the refractive index of the medium. These refractive index changes constitute a three-dimensional grating that diffracts part of the incident light beam, as shown below, causing a modulation in irradiance of both the directly transmitted beam and the diffracted beams. No polarizers are needed for this modulation.

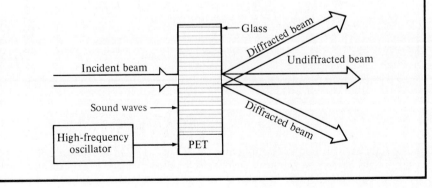

Q-switched pulses. The first of these is the *quarter-wave retardation mode*. Light within the cavity is passed through a linear polarizer mounted next to an electro-optic crystal. A voltage is applied to the crystal to produce a quarter waveplate (Section 2.7), converting the linearly polarized light to circularly polarized light. Upon reflection and return through the crystal, the light is again linearly polarized, but the plane of polarization is rotated 90° with respect to its initial direction. Since the polarizer prevents transmission at this rotated polarization direction, the crystal and polarizer together act as a closed shutter. To open the shutter, a negative pulse is applied to the crystal to reduce the net voltage across the crystal to zero, removing the polarization rotation. The voltage pulse is timed to occur at the peak inversion of the laser medium to form the Q-switched light pulse.

HALF-WAVE METHOD

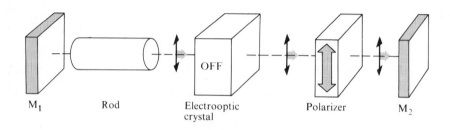

M₁   Rod   Electrooptic   Polarizer   M₂
                             crystal

(a) Voltage off:

Polarized wave from the rod travels ▥▶ down
the cavity and may stimulate atoms upon reflection.

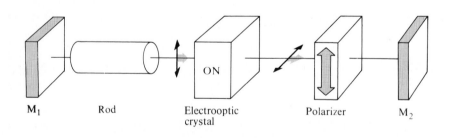

M₁   Rod   Electrooptic   Polarizer   M₂
                             crystal

(b) Voltage on:
Rotated polarization is absorbed by the polarizer.

Fig. 5.9 Half wave method of electrooptic Q-switching. (a) With no voltage applied to the crystal, the plane of polarization is unaffected. (b) When voltage is applied, the plane of polarization is rotated 90° upon passage through the crystal, and the light is absorbed by the polarizer. It is when the voltage is reduced to zero as in (a) that Q-switching occurs.

An alternative *half-wave retardation mode* differs from the quarter-wave method in that the crystal rotates the plane of polarization by 90° in a single pass, as shown in Fig. 5.9. With an appropriate dc voltage on the crystal, the polarizer acts as a closed shutter to the rotated light. When the voltage is reduced to zero, there is no rotation of the plane of polarization, and *Q*-switching occurs. This configuration is easier to align than the quarter-wave configuration, but has the disadvantage of requiring twice the operating voltage. In general, electrooptic shutters for *Q*-switching allow precise timing and sharp pulses, but these devices are not as rugged as the rotating prism.

Another *Q*-switching method relies on the *acoustooptic modulator* (again, see box). The modulator is located within the laser cavity. Application of a signal deflects some of the beam from the cavity, creating a high loss. As in the case of the electrooptic shutter method, the sound wave is shut off when the gain of the lasing medium is near its peak value, thereby forming a *Q*-switched pulse. Acousto-optic devices are most commonly used if the lasing medium is continuously pumped and repetitively *Q*-switched, as is frequently the case with neodymium lasers.

Passive *Q*-Switching

By relying entirely on the absorption characteristics of certain dyes, it is possible to *Q*-switch a laser without electronic circuits. "Passive" *Q*-switching can be obtained by placing in the laser cavity a material that exhibits an absorptivity that decreases with increasing irradiance, as shown in Fig. 5.10. An example of

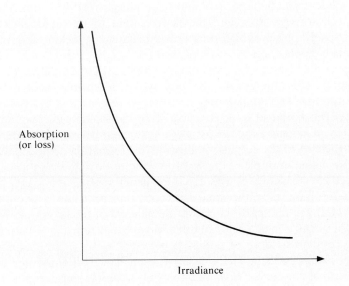

Fig. 5.10 Loss versus laser irradiance of a saturable absorber.

such a material is a saturable dye that possesses an absorption band at the lasing transition. At the beginning of the excitation flash, the dye is opaque due to the large number of unexcited molecules that can absorb the light. As in the case of the rotating prism, the low cavity Q prevents lasing action and allows a larger population inversion to be achieved than would otherwise occur. As the light irradiance in the cavity increases, more excited states of the dye are populated, until all possible excited states of the dye are filled. At this point, the dye can no longer absorb at the laser wavelength and is said to be "bleached." The abrupt reduction in cavity losses again causes Q-switching to occur. Passive Q-switching has the great advantage of being extremely simple to implement. The only equipment necessary is a small transparent cell containing the dye in a solvent inserted into the cavity between the lasing medium and one of the end mirrors. For ruby lasers, common dyes are cryptocyanine and vanadium pthalocyanine dissolved in organic liquids such as methanol or acetonitrile. Carbon dioxide lasers have been successfully Q-switched with sulfur hexafloride gas.

5.4 MODE LOCKING

An alternative technique for producing high-power, short-duration laser pulses is that of mode locking. A typical inhomogeneously broadened laser cavity may sustain the oscillation of many modes (Section 4.3). The output of such a laser, as a function of time, depends on the relative frequencies, phases, and amplitudes of these modes. In an ordinary laser, all these parameters are time-varying and the output fluctuates randomly. If, however, an external perturbation forces the various mode oscillations to maintain fixed phases relative to one another, the output becomes repetitive, and the laser is said to be "mode-locked." These regularly spaced pulses of high peak power make mode locking extremely attractive for many applications.

To generate high-power, short-duration pulses, a large number of modes must be oscillating. This follows from the same type of mathematical analysis that can demonstrate that an infinitely long duration wave is needed to produce a single frequency. The argument is the converse: To obtain a short-duration waveform (a pulse), many frequencies are needed. Although a real laser can operate with many mode frequencies in the output, we illustrate the formation of mode-locked pulses by considering an example of the superposition of three low-frequency waves. In Fig. 5.11, three cosine waves with equal frequency spacing, $\Delta v = c/2L$, are added to form a resultant wave, the square of which represents the laser output power. In Fig. 5.11(a), the phases of the waves are different at time $t = 0$; in Fig. 5.11(b), on the other hand, the phases of all three waves equal zero at $t = 0$. In the second case, the result is a series of pulses of period $T = 1/\Delta v$, the reciprocal of the frequency spacing. It is, in fact, a general result that if all phases are a constant, independent of the mode, then there is only a single pulse within the period

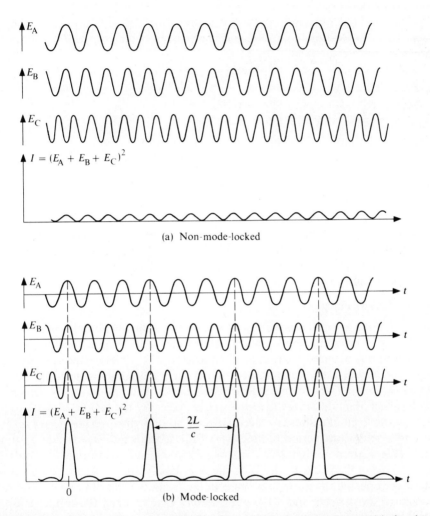

(a) Non-mode-locked

(b) Mode-locked

Fig. 5.11 Comparison of non-mode-locked and mode-locked outputs. In part (a), the phases are random and the instantaneous power is never large. In part (b), all the cosine waves have the same phase at $t = 0$. The narrow pulses are spaced $2L/c$ in time, the round-trip time for light in the cavity.

$T = 2L/c$. This corresponds to the round-trip transit time of a pulse within the laser cavity. The width of this pulse, τ, is approximately the inverse of the frequency spread of the laser output,

$$\tau = \frac{1}{\left(\dfrac{Nc}{2L}\right)} = \frac{2L}{Nc} \tag{5.2}$$

where N is the number of oscillating modes. We see that the ratio of the pulse spacing to the pulse width approximately equals the number of modes:

$$\frac{T}{\tau} = \frac{2L}{c} \cdot \frac{Nc}{2L} = N \tag{5.3}$$

Summarizing the above discussion, there are two conditions that must be fulfilled to obtain narrow mode-locked pulses. (1) There must be a large number of oscillating modes, and (2) they must be forced into definite phase relationships. The former condition depends on a broad laser transition and a long cavity to give small intermode separation. The latter condition can be fulfilled by either active or passive mode locking.

Active Mode Locking

We have stated above that mode locking is accomplished by forcing the longitudinal modes to maintain fixed phase relationships. This can be accomplished by varying the loss of the laser cavity at a frequency equal to the intermode separation, $c/2L$. The loss may be accomplished by the same methods discussed in the preceding section on Q-switching—electrooptic and acoustooptic shutters. To visualize the effect of modulating the losses at $c/2L$, consider the loss in the laser cavity to be in the form of a shutter placed adjacent to one mirror. The shutter is open for only a very short period of time every $2L/c$ seconds and is closed at all other times. (We again note that $2L/c$ corresponds to the round-trip time of a pulse of light bouncing back and forth inside the laser cavity and is the reciprocal of the intermode spacing.) If we have a pulse of light that is exactly as long in time as the shutter stays open, and if the pulse arrives exactly when the shutter is open, it will be unaffected by the presence of the shutter. If a portion of the pulse arrives when the shutter is closed, it will be eliminated. The net result of the periodic modulation is a single pulse that bounces back and forth between the mirrors and is of duration approximately equal to $2L/Nc$. Each time the pulse reflects from the output mirror, a portion is emitted, yielding a periodic pulse train spaced in time by $2L/c$.

Passive Mode Locking: Saturable Absorption

Just as a saturable absorber can be used to produce a simple, effective Q-switch, it also can be used to mode-lock. In the initial stages of formation of the mode-locked pulse, the laser medium emits spontaneous emission, or noise, which consists of incoherent fluctuations in the cavity energy density. Some of these fluctuations are of extremely short duration. These initial fluctuations are coherently amplified by the laser medium and grow in intensity. If an intense pulse of light passes through the saturable absorber inside the laser cavity, the low-power tails of the pulse are attenuated because of the absorption of the dye molecules. The high-power peak of the pulse, however, is transmitted because the dye is bleached. In order for this to occur, the saturable absorber must recover in a time short compared to the duration of the pulse. Because of this nonlinear bleaching mechanism, the shortest and most intense fluctuations grow at the expense of the weaker ones. With careful positioning of the dye cell in the laser cavity and adjustment of the dye concentration, the initial laser noise can grow into a narrow pulse that shuttles back and forth in the cavity, creating the desired periodic pulse train. Saturable absorbers provide a simple, inexpensive, and rugged method of mode locking high-power, pulsed lasers. When a saturable absorber is used to mode-lock, the laser is simultaneously mode-locked and Q-switched. The result is a series of narrow, mode-locked pulses contained in an envelope that may be as long as several hundred nanoseconds. The individual mode-locked pulses are extremely narrow, on the order of tens of picoseconds. Because of the extremely short duration of these pulses, the peak powers in a mode-locked laser beam can be enormous.

PROBLEMS

5.1   If the full linewidth at half maximum of a helium-neon laser is 1.5 GHz (1.5×10^9 Hz), and if the cavity mode must be within this linewidth for laser action to occur, what must be the length of the laser cavity to obtain single-mode operation?

5.2   If a cavity mode burns a hole in the side of a gain-versus-velocity curve for the 633 nm neon transition (Fig. 5.5a) at the half-maximum, what is the velocity of the atoms partaking in the hole burning, given that the full linewidth at half maximum is 15 GHz?

5.3   Calculate the approximate peak power in a Q-switched ruby laser pulse ($\lambda = 694.3$ nm) if the pulse has an energy of 4 joules and a duration of 40 nsec. (Assume a rectangular pulse shape.) If there are a total of 5×10^{19} chromium ions, what is the percentage population inversion required to achieve a 3-joule output if each ion in the upper state emits only once during the pulse duration? The percentage population inversion is defined as

$$\frac{N_1 - N_0}{N_1 + N_0} \times 100\%$$

where N_1 is the population of the upper state of the laser transition, and N_0 is that of the lower.

5.4   Calculate the mode-locked pulsewidth for a Nd:glass laser, assuming that the laser cavity is 50 cm long and that there are 1000 active longitudinal modes. How far apart are the mode-locked pulses spaced in time?

5.5   It is proposed to measure the distance from the earth to the moon by firing a Q-switched ruby laser ($\lambda = 694.3$ nm) of pulsewidth 50 nsec and observing the time taken for the reflected light to return to earth. The light is reflected by a corner cube array placed on the moon's surface. The corner cube array directs the laser beam back on itself regardless of the direction from which the beam comes. The laser beam is passed through a telescope to reduce the beam divergence. (The beam divergence is reduced by the ratio of the diameter of the telescope aperture to that of the laser beam diameter.)

a)   Assuming that the moon-to-earth distance is approximately 3.8×10^8 m, and that the diameter of the laser beam is 1 cm with a full angle beam divergence of 10^{-4} rad, calculate the approximate spot size on the moon using a telescope with aperture 300 cm (comparable to that of the telescope at Lick Observatory in California, which was used in some lunar ranging experiments).

b)   What fraction of the light would be reflected by a square corner cube array 46 cm on a side?

c)   What fraction of the initial beam would be received by the transmitting telescope, assuming that the divergence of the reflected light from the corner cube array is determined by the diameter of a single corner cube (3.8 cm) in the array?

d)   About how long does it take the light to complete a round trip from the earth to the moon?

e)   Assuming that the uncertainty in the measurement is approximately equal to the distance traveled in the time of one laser pulsewidth, to what accuracy can the measurement be performed using this method?

REFERENCES

5.1   B. A. Lengyel (1971), *Lasers* (2nd edition). New York: Wiley.

5.2   A. Maitland and M. H. Dunn (1970), *Laser Physics*. New York: American Elsevier.

5.3   R. H. Pantell and H. E. Putoff (1969), *Fundamentals of Quantum Electronics*. New York: Wiley. Excellent discussion of the rate equations without the prerequisite of quantum mechanics. The rest of the book does require quantum theory for understanding.

5.4   Editors of *Scientific American* (1969), *Lasers and Light*. San Francisco: W. H. Freeman. Reprints of *Scientific American* articles with introductions to major sections by A. L. Schawlow. Nonmathematical readings. The laser-related articles occupy the final third of the book. Some of the sections are now dated.

5.5   A. E. Siegman (1971), *Introduction to Lasers and Masers*. New York: McGraw-Hill.

5.6   O. Svelto (1976), *Principles of Lasers*. New York: Plenum.

5.7   A. Yariv (1976), *Introduction to Optical Electronics* (2nd edition). New York: Holt, Reinhart and Winston.

5.8   A. Yariv (1975), *Quantum Electronics* (2nd edition). New York: Wiley.

A Laser Exposition

Development of new laser devices and materials has proceeded rapidly since 1958, when A. L. Schawlow and C. H. Townes proposed that the principle of amplification of radiation by stimulated emission could be extended to the optical region of the spectrum. The first laser was constructed in 1960 by T. H. Maiman at the Hughes Aircraft Corporation Research Laboratories. It was operated on a pulsed basis and employed a crystal of pink ruby as the active medium. Later that same year, the first CW* laser was constructed by A. Javan at the Bell Telephone Laboratories. This laser used a mixture of helium and neon gas as the active medium, excited by a radio-frequency electric field. Semiconductor lasers, analyzed theoretically by N. G. Basov and coworkers in Russia, were developed simultaneously in 1962 by several American laboratories (IBM, General Electric, and Lincoln Laboratories). Since that time, there has been a rapid increase in the types of lasers and in the materials in which lasing has been demonstrated to occur.

Obviously, it is not possible to cover such a rapidly developing technology in great detail. In this chapter, we shall concentrate on a description of the most common types of lasers—gas, doped insulator, semiconductor, and dye—these four categories representing the great majority of lasers in use today. Each laser system discussed satisfies in some way the three fundamental requirements of a laser: an active medium, a population inversion, and some form of optical feedback. The active medium is associated with the *laser material* (e.g., the helium-neon gas mixture, ruby crystal, or gallium arsenide semiconductor). The population inversion is achieved through a *pump excitation* (gas discharge, flashlamp, high injection current). The third requirement, feedback, is brought about by the

---

* The notation CW is one of many pieces of laser jargon. It means "continuous wave," implying uninterrupted, as opposed to pulsed, output from the laser.

enclosure of the medium in an *optical cavity* (dielectric mirrors, polished ends of a crystal rod, cleaved crystal faces). (As we pointed out in Section 3.3, this requirement is not fulfilled in all cases. Some examples will be given.) Other components may be added to modify the output or to select a particular lasing frequency, but they are not necessary for lasing to occur. When reviewing the material included in this chapter, the reader may find it useful to identify the various laser components and the requirements that they satisfy.

6.1 GAS LASERS

Gas lasers are the workhorses of the laser industry. They range from the powerful industrial carbon dioxide units to the ubiquitous helium-neon lasers of modest powers. They can be operated continuously or on a pulsed basis; their output frequencies range from the ultraviolet to the infrared. Depending on the nature of the active medium, three types of gas lasers can be distinguished: atomic, ionic, and molecular. Although several different excitation mechanisms have been employed for pumping them, most gas lasers are excited by means of an electric discharge. Electrons in the discharge are accelerated by the electric field between a pair of electrodes. As the electrons collide with the atoms, ions, or molecules of the active medium, they induce transitions to higher energy states. With sufficient pumping, a population inversion is created.

Atomic Lasers

The principal example of a laser that utilizes a transition between energy levels of nonionized atoms is the helium-neon laser. The lasing medium is a mixture of ten parts helium to one part neon. Only the energy levels of the neon atom are directly involved in the laser transition; the helium gas is present to provide an efficient excitation mechanism for the neon atoms. Most helium-neon lasers are excited by a direct-current (dc) discharge, created by placing a high voltage across a gas-filled space. The helium atoms are easily excited by electron impact to any one of several low-lying metastable energy states.

The neon atom, having six more electrons than the helium atom, has an extremely complicated distribution of excited states. Two of its higher energy states have almost exactly the same energy as two of the metastable helium states. With the energy match so close, a collision between a helium atom and a neon atom can result in the efficient transfer of energy from the metastable helium atom to the unexcited neon atom. The helium atom returns to its ground state upon excitation of the neon atom into its excited state. A collision that results in this type of energy transfer is called a *resonant collision*. A diagram of the energy states for helium and neon is shown in Fig. 6.1. The helium excited states are identified by combinations of letters and numbers, 2^1S and 2^3S, which specify the total angular momentum and spin of the two electrons in the excited atom. The neon excited states are identified by the quantum numbers of the single excited state electron. (The other nine electrons in the neon atom retain their

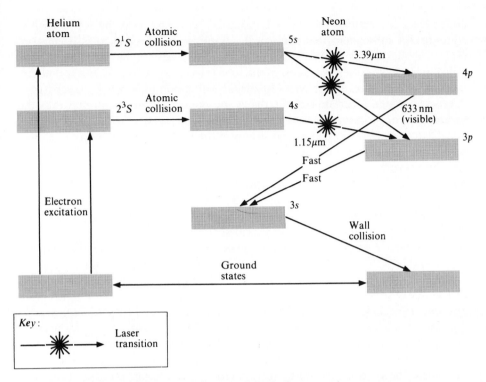

Fig. 6.1 Energy-level diagram of the helium-neon laser system.

ground-state quantum numbers.) As noted earlier, these numbers determine the probability of a given transition. For example, a quantum mechanical calculation shows that the transitions between the neon s states (e.g., $5s \rightarrow 4s$), are forbidden. The helium atoms are excited to metastable levels 2^1S and 2^3S by direct electron impact. The helium atoms then collide with the unexcited neon atoms and the neon atoms are raised to the $5s$ and $4s$ states. These states have longer lifetimes than the lower-energy $4p$ and $3p$ states. Thus the $5s$ and $4s$ states are pumped by the metastable helium atoms, while the $4p$ and $3p$ states are depleted because of their short lifetimes. An inversion of the populations between the s and p states results, and amplification by stimulated emission occurs. The population inversion is increased substantially if the excited neon atoms are allowed to collide with the walls of the chamber confining the discharge. The collisions allow non-radiative transitions* to take place between the $3s$ and ground states of the neon

---

* A transition in which deexcitation is not accompanied by an emission of radiation is said to be *nonradiative*. The stored energy is given up by the atom to its surroundings as thermal motion; or in the case of a molecule, the energy may be converted to molecular vibrations.

atom; these transitions prevent a buildup of neon atoms in the lower excited states and a subsequent reduction in the population inversion. It is because of this particular depopulation mechanism that one cannot increase the output power of a HeNe laser by increasing the tube cross section. The reason is that any increase in the radius of the *bore*, the cylindrical region to which the discharge is confined, beyond a certain value reduces the population inversion and thus the overall gain of the laser.

There are many more laser transitions in the HeNe laser than we have shown in Fig. 6.1. Each of the energy states of neon (shown as bars in the diagram) is split into several sublevels. Each sublevel can serve as the initial or terminal level for several different laser transitions, producing the 130-plus stimulated emission lines that have been observed in neon. All of the lasing lines can be produced in a discharge of pure neon. However, the output of many lines is greatly increased by the resonant collision pumping described.

Both the 633 nm and the 3.39 μm transitions start with the same upper energy state (5s). The 3.39 μm (infrared) transition has a much higher gain than the 633 nm (visible red) transition and can deplete the 5s level, reducing or eliminating completely the visible output of the laser. Several techniques can be employed to discriminate against the 3.39 μm transition and to encourage the 633 nm transition. In the method most commonly employed, the laser mirrors are designed to be highly reflective at 633 nm but highly transmissive at 3.39 μm. The round-trip gain at the visible wave transition can then be satisfactorily high, while at the same time the gain for the infrared transition never reaches threshold. Another technique consists of placing small magnets along the length of the laser tube, thereby creating an inhomogeneous magnetic field. The magnetic field produces a splitting (Zeeman splitting) of certain spectral lines into several components. It is possible to show that the gain per unit length at the lasing transition is inversely proportional to the linewidth. The Zeeman splitting broadens the infrared 3.39 μm laser line more than the visible line, decreasing its gain, so that the visible transition is favored.

The basic construction of a helium-neon laser is relatively simple. The essential elements are the discharge tube containing the gas, the power supply, and the mirrors. A mass-produced HeNe laser is shown in Fig. 6.2. Figure 6.3 is a simplified schematic diagram for one kind of dc power supply. A large resistor, called a *ballast resistor*, is placed in series with the discharge tube. The ballast resistor is required because the discharge tube exhibits negative dynamic resistance; with no ballast resistor present, once a discharge current is initiated, the current increases as the voltage across the tube *drops*. The ballast resistor limits the current, protecting the power supply and stabilizing operation of the tube. Sometimes a saturable core transformer, of the kind employed in neon signs, is used for the same purpose. The output voltage of the transformer is initially high, sufficient to initiate discharge. At high currents, however, the transformer saturates, automatically reducing the output voltage and limiting the discharge current.

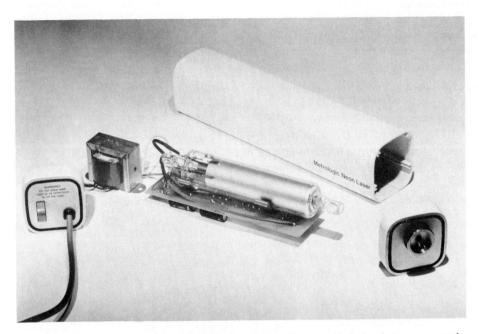

Fig. 6.2 Mass-produced helium-neon laser, consisting of an extruded aluminum case, a voltage transformer, a voltage multiplier and rectifying circuit on a printed circuit board, and a laser tube. (Courtesy of Metrologic Instruments, Inc.)

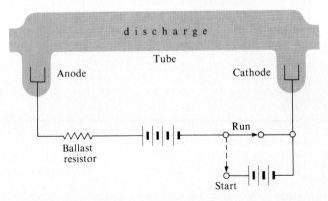

Fig. 6.3 Simplified electrical circuit for a gas laser. A larger voltage is needed to start the discharge than to maintain it, so a high-voltage pulse is applied to the gas when the laser is turned on. The ballast resistor serves to limit the current once the discharge is initiated.

A typical low-output-power HeNe laser operates with a dc voltage of about 1600 volts and draws a current of 5 milliamperes. This amounts to an input power of about 8 watts. For a laser output power of about 1.5 milliwatts, the overall efficiency is about 0.02 percent. Excitation by radio frequency electric fields is also possible, but the overall laser efficiency is not as high as with dc excitation. One can also apply an ac voltage across the tube; the result is a laser with 60 Hz modulation, a glorified neon sign!

The optical cavity consists of two mirrors, whose radii of curvature and separation constitute a stable optical resonator configuration. An *internal mirror arrangement* has two mirrors cemented directly to the ends of the discharge tube.

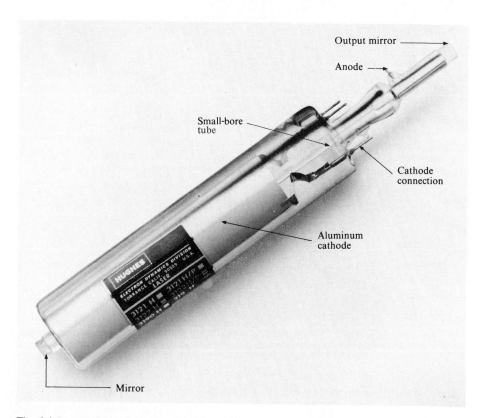

Fig. 6.4 Laser tube with internal mirrors. The large tube is fitted internally with a small-bore tube (about 3 mm inside diameter). The bore confines the discharge to the region between the mirrors and provides the surface for wall collisions needed to maintain the population inversion in the HeNe system. (Courtesy of Hughes Aircraft Company, Electron Dynamics Division.)

Alignment of the mirrors, which is critical to the achievement of a high power output, is performed before the cement sets. This method has the advantage that the laser is prealigned and cannot be misaligned without bending or breaking the tube. An example of such a tube is shown in Fig. 6.4. A second arrangement consists of the tube placed between two mirrors, external to the laser discharge tube, which are mounted on the laser frame. The tube is sealed off with Brewster windows, which allow 100-percent transmission of the laser light for one polarized direction and produce a highly polarized output (see Section 2.7). The *external mirror arrangement* is more complicated and expensive than the internal mirror arrangement, because in addition to protection against dust and dirt inside the cavity, one must have excellent stability of the mirror mounts and the tube supports. The advantage of the external mirror arrangement is that frequency-selective and light-switching devices can be inserted into the cavity. Also, mirrors can be changed for different frequency ranges and coherence requirements.

The mirrors used in these and many other types of lasers are produced using techniques that have been devised in response to laser design requirements. Metallic mirrors have such high losses due to absorption that they do not allow oscillation to be sustained in any but the highest gain lasers. Typical laser mirrors are made up of alternating layers of high- and low- refractive index dielectric materials, each layer a specific fraction of a wavelength thick. The resulting interference between waves reflected from and transmitted through these layers results in a highly reflective mirror with virtually no absorption or scattering losses. The high reflectivity of these *dielectric mirrors* extends only over a limited portion of the spectrum. For the helium-neon laser, the reflectivity of the mirrors is high at 633 nm, as required, but often falls off in the blue region. As a result, one sometimes notices a halo of blue light around the red output of a HeNe laser—some of the shorter wavelength light from the glow discharge has been transmitted.

Most helium-neon lasers have power outputs of less than 50 mW, the great majority with powers in the 0.5–5.0 mW range. Because of their low power output, the small HeNe lasers are often thought to represent no hazard to the user. It should be noted, however, that the collimated beam of such a laser can be focused by the lens of the eye to a spot of potentially damaging power density. Looking down the bore of *any* laser is quite dangerous. Relatively little physical damage to the fovea, the area of sharpest vision, can result in substantial loss of sight.

Noble Gas Ion Lasers

The most powerful visible lasers that operate continously are the noble gas ion lasers. Outputs of several watts in a single line can be obtained from some of these lasers. Although this may seem unimpressive when compared to the 60–100 watt output of a standard light bulb, keep in mind that the light bulb's output is spread across the entire visible spectrum and extends into the infrared. The two watts of an argon laser line, on the other hand, may be contained within a line-width of a fraction of a nanometer.

The rare, or noble, gases—helium, neon, argon, xenon, and krypton—have filled electron shells and, consequently, do not normally combine chemically with other elements. Although their chemical usefulness is limited, they have exhibited remarkable ability to support stimulated emission. As we have stated, in neon alone there are over 130 possible laser transitions, ranging from the yellow into the infrared portions of the spectrum. However, with the exception of neon, none of the neutral rare gases is very useful as a practical laser medium. The other noble gases, when used as laser media, must be ionized by electron collisions. The resulting ion is excited by further electron collisions and an inverted population of the ion energy levels is achieved. To ionize the noble gas atoms, a high current (of the order of 15 to 50 amps) must be passed through the discharge tube. By forcing the discharge current through a tube of small bore, the current density can be made as high as 10^3 amps/cm^2. To increase the current density further and to decrease the number of ions and electrons colliding with the walls, a magnetic field is applied to the tube by placing it inside a large solenoid. The field causes the electrons to spiral about the magnetic lines of force that are parallel to the tube axis. Because of the high current involved, the cathode must be an excellent electron emitter. Any contaminants in the tube can "poison" it, reducing the emission and thereby the current and the laser output.

The spectroscopy of the noble gas ions is quite complicated and not as well understood as that of the neutral atoms. The two strongest laser lines in the argon ion laser occur at 488.0 nm and 514.5 nm, in the blue and green portions of the spectrum, respectively. The argon laser, because of these two strong lines, is the most common noble gas ion laser. Krypton is also frequently used, because its transition lines, although weaker than argon, are spread across the entire visible spectrum. Both of these lasers have ultraviolet lines that are used to pump a tunable dye laser (see Section 6.4).

The tube design of an argon ion laser is much more complicated than that of a helium-neon laser. The high current through the small bore demands refractory materials that can withstand the intense heating caused by the highly excited gas. Several materials, including quartz, graphite, and beryllium oxide (BeO), have been used to fashion the bore. The discharge is initiated by a high-voltage trigger pulse applied across the tube. During operation, the high current causes a pumping of the argon ions toward the cathode and the electrons toward the anode. Because the mobility of the ions is much less than the mobility of the electrons, there is a tendency for the ions to pile up at the cathode. There they are neutralized and diffuse slowly back into the discharge. If nothing is done to counteract this distribution of argon ions near the cathode, the discharge is extinguished. For this reason, a gas return path is provided between anode and cathode by introducing staggered off-axis holes in the bore material, as shown in Fig. 6.5. This return path permits diffusion of the ions back to the anode to equalize the pressure caused by the pumping, yet it is sufficiently complicated that the discharge is confined to the direct bore path. To dissipate the large amount of heat generated in the bore, most ion laser tubes are water cooled. Although some lasers have separate heat exchangers, most use filtered tap water. Simple pulsed versions of argon ion lasers are available. Because the duty cycle ("on" time divided by the time between pulses) is low, the heat energy generated is small, and usually only convective cooling is needed. The average power output is of the order of milliwatts, though the peak powers can be as high as several watts. Pulsewidths are approximately 5 to 50 microseconds, with repetition rates up to 60 pulses per second. These low-power, pulsed ion lasers are generally much less expensive than their CW counterparts. Most ion lasers tend to lase in several strong lines simultaneously, and it is frequently necessary to select a single line. This is accomplished by the insertion of a wavelength-dispersing element into the cavity, e.g., a grating or a prism, as described in Section 5.1. For maximum power in the beam (as is the case for pumping dye lasers), it may be desirable to remove the prism and let many lines lase. For certain applications, such as color TV and color holography, several gases are introduced into the tube and all lase simultaneously to yield a "white light" output. For example, a mixture of argon and krypton gas can produce strong blue, green, yellow, and red laser lines.

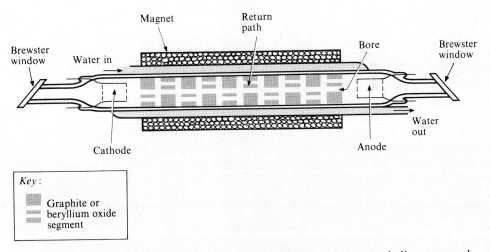

Fig. 6.5 Construction of an ion laser tube. The water jacket and magnet windings surround the tube. The return path permits diffusion of ions back to the anode to equalize the pressure caused by a pileup of neutralized ions at the cathode.

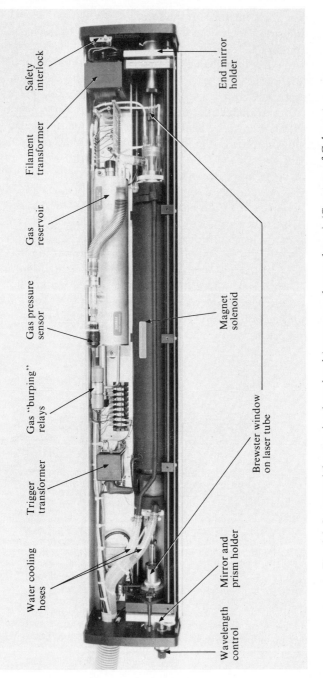

Water cooling
hoses

Trigger
transformer

Gas "burping"
relays

Gas pressure
sensor

Gas
reservoir

Filament
transformer

Safety
interlock

End mirror
holder

Magnet
solenoid

Brewster window
on laser tube

Mirror and
prism holder

Wavelength
control

Fig. 6.6 Commercial argon ion laser head (power supply not shown). (Courtesy of Coherent Radiation, Inc.)

A photograph of a commercial argon ion laser with its top cover removed is shown in Fig. 6.6. The tube is filled and sealed at the correct gas pressure when the laser is manufactured; however, the argon gas is gradually absorbed into the graphite that forms the bore of the laser tube. To maintain the correct pressure, an auxiliary gas "burping" system is included to replenish the tube. Tubes with a beryllium oxide bore are manufactured as sealed units, since they do not suffer the problem of gas absorption. The trigger transformer provides the high voltage necessary to start the discharge. The filament transformer provides the low voltage that is used to heat the cathode to obtain electron emission. Because of the burping system, tube life is generally limited by the life of the cathode. The large discharge current is controlled by a bank of transistors arranged in parallel inside a large external power supply (not shown in the photograph). Each transistor regulates a fraction of the total tube current. Because of the high power requirements, most argon ion lasers operate on 208-volt, three-phase electrical lines.

All of the hazards attendant in a helium-neon laser are present in argon and krypton ion lasers. The hazard is greater in the latter, however, since the eye is more sensitive to damage at the output wavelengths of these lasers and the output power levels are generally greater. Safety goggles that absorb the output wavelengths may be necessary with many of these devices. However, one should ascertain the location and direction of the beam before approaching the area. Skin injuries can occur at high power levels. Even at moderate power levels, an unfocused argon ion laser beam can burn holes in clothing. Several scientists have lost their ties while trying to obtain a better view of their research.

Metal Vapor Lasers: HeCd and HeSe Lasers

The first metal vapor laser, using cesium vapor as the active medium, was operated in the pulsed mode and had an output in the infrared. Later research led to the development of continuous-wave metal vapor lasers that could operate in the visible portion of the spectrum. Two of these lasers, the helium-cadmium and helium-selenium lasers, combine the features of the two types of lasers we have just discussed, the helium-neon atomic system and the noble gas ion system. The energy-level diagrams for both of these lasers are shown in Fig. 6.7. In metal vapor lasers, the metallic atoms are heated to form a vapor in the discharge tube, which also contains some helium gas. In the helium-cadmium laser, metastable excited helium atoms created in the discharge collide with the cadmium atoms, and a transfer of energy occurs. In this transfer, the cadmium atom is ionized and the helium atom is returned to its ground state. Any energy difference between the two states is carried away by the freed electron. Such a process is called a *Penning ionization*.* The cadmium ions are left in an excited state, from which

---

* The Penning ionization is named after F. M. Penning, a Dutch physicist who made many significant contributions to science through his research on gas discharges.

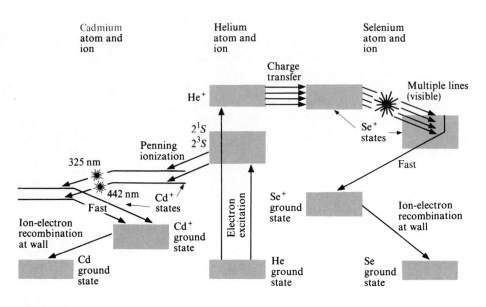

Fig. 6.7 Energy-level diagrams for two metal vapor lasers using helium pumping. On the left is the diagram for the HeCd laser; on the right is that for the HeSe laser. Charge transfer populates the selenium ion levels; Penning ionization populates the cadmium ion levels. The energy mismatch in the Penning ionization is taken up by the electron released upon ionization.

the laser transition occurs. A population inversion is maintained, because the ions decay faster from the lower levels than from the higher levels. Neutralization of the cadmium ion occurs at the walls of the tube.

The excitation of the helium-selenium laser differs slightly from that of the helium-cadmium laser in that helium ions, instead of helium atoms, collide with the neutral selenium atoms. An electron is transferred from the selenium atom to the helium ion, leaving the resulting helium atom neutral and the selenium atom ionized. This process is called a *charge-transfer* process. The ionized states produced by this process are the upper levels of a number of laser transitions. Again, it is electron-ion recombination at the tube walls that returns the selenium to its neutral ground state.

The construction of metal vapor lasers differs from that of other atomic lasers in that provisions must be made to contain the metal, to heat it to a vapor, and to distribute the vapor within the gas discharge. The solid metal is contained in one end of the tube near the anode and is heated to vaporize a sufficient quantity of metal into the discharge region. When the metal enters the discharge, some of the metal atoms are positively ionized and are attracted toward the cathode. The metallic ions drift along the length of the tube and form the active laser medium. In the cathode region, where there is no discharge, the temperature is lower and the metal vapor condenses on the tube. This process of continuously transferring the metal from anode to cathode is termed *cataphoresis pumping*. When all of the metal originally at the anode end of the tube is used up, laser action ceases. This may be quite a long time; the rate of cadmium usage in a small helium-cadmium laser may be lower than one gram per 700 hours of operation. Techniques that permit reuse of the metal have been developed, such as the provision of a return path for the atoms through segmented bores, as in the argon laser. Another scheme uses an additional anode built next to the first cathode, and an additional cathode next to the first anode. Each time the laser is turned on, an alternate anode-cathode pair is used, and current and metal flow proceeds in the opposite direction.

Most commercial HeCd lasers have output powers of 5 to 15 mW at $\lambda = 442$ nm (blue) and about 1.5 mW at $\lambda = 325$ nm (ultraviolet). Different sets of mirrors are required for operation at these two wavelengths. If the tubes have internal mirrors, as is often the case, the output is restricted to one of the two wavelengths. The cost per milliwatt of output power is approximately equal to that of helium-neon lasers. Because these lasers are air cooled and plug into an ordinary electrical outlet, their blue output is a low-power alternative to that of the more costly argon ion laser. The HeSe laser output ranges from 460 nm to 650 nm. Again, more than one set of mirrors is required to obtain all fifteen possible lasing transitions. Usually, one mirror set is used for lines between 460 and 530 nm, and another mirror set is used for lines between 550 and 650 nm. The total output for all lines in the blue-green region of the system is approximately 10 to 20 mW. Recently, a modification of the HeCd laser tube, in which the bore of the laser is in an extended cathode region of the discharge, has produced a number of new lines in the green, yellow, and red areas, making it a multicolor source like the HeSe laser. The powers for several lines can be balanced to produce an output beam that is "white." (The speckles, however, are colored!)

Molecular Lasers

From the standpoint of potential industrial applications, the carbon dioxide laser unquestionably ranks first. The CO_2 laser offers both high power and high efficiency at an infrared wavelength. Carbon dioxide lasers have been used to weld metals, cut ceramics, and perform many other materials-processing tasks. Some of them will be discussed in Chapter 9. The CO_2 laser is the most important example of the class of lasers referred to as *molecular lasers*. Thus far in our discussion, the energy levels of interest for laser transitions have been electronic energy levels of an atom or an ion. Molecules have a more complicated structure and have energy levels that correspond to rotating or vibrating motions of the entire molecular structure.

The carbon dioxide molecule, composed of two oxygen atoms and a carbon atom between them, undergoes three different types of vibrational oscillation, as shown in Fig. 6.8. These three fundamental vibrational configurations are called *vibrational modes* (not to be confused with the modes of the laser cavity). According to quantum theory, the energy of oscillation of a molecule in any one mode can have only discrete values, just as the energy of an electron in an atom is quantized. The discrete values are all integer multiples of some fundamental value. At any one time, a carbon dioxide molecule can be vibrating in a linear combination of the three fundamental modes. The energy state of the molecule can then be represented by three numbers, $(i\ j\ k)$. These numbers represent the amount of energy, or number of energy quanta, associated with each mode. For example, the number (002) next to the highest energy level shown in Fig. 6.9 means that a molecule in this energy state is in the pure asymmetric stretch mode with two units of energy (i.e., no units of energy associated with the symmetric stretch or bending mode). In addition to vibrational states, rotational states, associated with rotation of the molecule about the center of mass, are also possible. The energies associated with the rotational states are generally small compared to those of the vibrational states, and are observed as splittings of the vibrational levels into a number of much finer sublevels. The separations between vibrational-rotational states are usually much smaller on the energy scale than separations

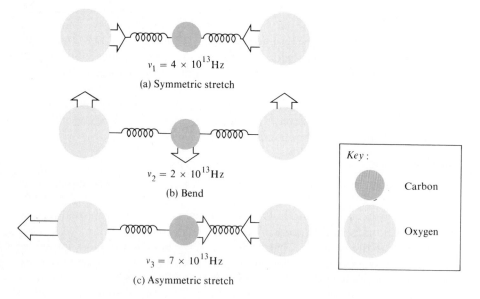

$v_1 = 4 \times 10^{13}$Hz

(a) Symmetric stretch

$v_2 = 2 \times 10^{13}$Hz

(b) Bend

$v_3 = 7 \times 10^{13}$Hz

(c) Asymmetric stretch

Key:

Carbon

Oxygen

Fig. 6.8 Vibrational modes of the CO_2 molecule.

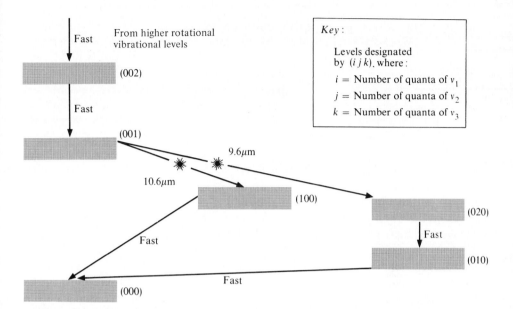

Fig. 6.9 Energy-level diagram for the CO_2 laser. Pumping occurs to the higher energy levels (not drawn to scale). Bands shown contain numerous discrete rotational levels.

between electronic states. The radiation associated with the energy difference between electronic transitions is usually visible or ultraviolet, whereas the vibrational-rotational transitions are in the near and far infrared. For this reason, most molecular lasers have infrared outputs.

The various low-lying energy levels of the CO_2 molecule that are responsible for the laser transitions are shown in Fig. 6.9. Each group of lines represents a different vibrational energy level; each individual line represents a different rotational energy level. In the CO_2 laser, molecules are pumped from the ground state to higher energy states (out of the diagram) from which they trickle back by radiative and nonradiative processes to state (001), which is metastable. With sufficient pumping, a population inversion is produced between the (001) state and the (100) and (020) states. If the losses in the laser cavity are sufficiently low, laser oscillations begin. The strongest line of the CO_2 laser is at a wavelength of 10.6 μm, in the infrared. A weaker line at 9.6 μm competes with the 10.6 μm line for the available excited molecules. For improved laser output, nitrogen and helium are generally added to the gas mixture (approximately 10 percent CO_2, 40 percent N_2, and 50 percent He). The nitrogen in the CO_2 discharge takes the role that helium plays in the helium-neon and metal vapor lasers; excited nitrogen molecules transfer energy to the CO_2 molecules by resonant collisions. The helium serves to increase the laser efficiency by speeding up the transition from the (100) energy level (the receiving end of the 10.6 μm laser transition) to the ground level, thereby maintaining a large population inversion.

Carbon dioxide lasers are capable of producing tremendous amounts of output power, primarily because of the high efficiency of the 10.6 μm transition. A well-constructed system can achieve an efficiency of about 30 percent, as compared to less than 0.02 percent for helium-neon lasers. Gigawatts of peak power have been obtained in short nanosecond-duration pulses. The principal difference between the CO_2 laser and other gas lasers we have discussed is that the optics must be coated or made of special materials so that they are reflective or transmissive in the infrared. If the cavity has external mirrors, the plasma tube usually has Brewster-angle windows fabricated from germanium, cadmium sulfide, or, in the case of high power systems, from sodium chloride or potassium bromide. These materials are transparent at 10.6 μm, a region where most other materials, including glass, are opaque. Although the latter two materials are sufficiently strong mechanically to serve well as Brewster windows, they are hygroscopic, that is, they absorb moisture from the atmosphere, and their optical surfaces take on a milky appearance if they are left unprotected for extended periods of time. The optical resonator itself is provided with a pair of long radius-of-curvature mirrors, with multilayer dielectric reflective coatings. The output mirror can be made of germanium or gallium arsenide, both of which if cooled have low loss at 10.6 μm. In some cases, a diffraction grating mounted on a piezoelectric transducer is used in place of the high-reflectivity mirror. The grating permits tuning of the laser output over twenty distinct lines within either of two major bands at 9.6 μm and 10.6 μm. The output reflector can be an uncoated gallium arsenide flat.

The power supply for a CO_2 laser must provide a sufficiently high voltage to maintain a discharge with cavity pressures of 10 mm of mercury or more. This voltage, which is about 8 kV per meter of discharge, constitutes a major hazard if not carefully shielded. The invisible output, with power levels approaching 100 watts for a laser one meter long, is in a beam the diameter of a pencil and is capable of inflicting severe burns on an unwary experimenter. A "Danger-Laser" sign is definitely appropriate.

The 10.6 μm spectral region of the laser output presents some difficulties in potential applications, because convenient detection methods have not been developed in that range. The available detectors are insensitive and expensive compared to those available at visible wavelengths, and generally require cooling to low temperatures. Nevertheless, because of its power and efficiency, the CO_2 laser is of great practical importance. With even a few tens of watts of power, a properly focused CO_2 laser beam can easily burn through a wooden plank, an asbestos sheet, or a metal plate. The power output of a CO_2 laser is approximately proportional to the active length of the laser. In attempts to obtain greater output power, researchers have built CO_2 lasers tens of meters long with CW output powers ranging to the tens of kilowatts.

TEA and Gas Dynamic Lasers

Two variations of the CO_2 laser configuration are useful for high-power applications. The first variation, frequently used in industrial applications, is called the TEA laser, TEA designating Transverse Excitation Atmospheric. Though not the case with all gas lasers, in the CO_2 laser the output power can be increased simply by increasing the pressure of the CO_2. Thus, if such a laser can be made to operate at atmospheric pressure or even above, substantially greater output power should be obtainable from a laser of given length. The difficulty with this approach is that approximately 12 kV per cm is necessary to initiate and maintain a discharge in the gas at atmospheric pressure. For a longitudinally excited laser one meter long, a monstrous high-voltage power supply would be necessary. The trick developed to overcome this difficulty is suggested in the first two letters of the TEA acronym—transverse excitation.

The discharge is arranged to take place at a number of points in a direction transverse to the laser cavity rather than longitudinally, as shown in Fig. 6.10. Since this arrangement results in a discharge pathlength of about a centimeter instead of the typical meter or more, the breakdown voltage needed is thus comparatively modest. With this arrangement, gigawatts of peak power can be obtained in very short pulses. Each individual discharge cathode is equipped with its own ballast resistor (in most schemes). Were this not the case, the negative dynamic resistance characteristics of the gas would lead quickly to a single discharge current path in the region of least resistance. Proper placement of the cathode-anode pairs is important, because a nonuniform discharge distribution leads to undesired output mode patterns. In the straight-line pattern shown in the figure, the laser tends to lase in an oval transverse mode pattern, because the

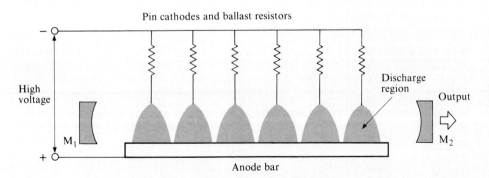

Fig. 6.10 TEA laser. The discharge occurs perpendicular to the laser cavity.

discharge is elongated in the direction of the pins. A round beam pattern is obtained if the cathode pins are arranged in a helix. In some high-power commercial CO_2 TEA lasers, a distributed cathode made of a screen, shaped somewhat like the top of an elongated loaf of bread, is used. The screen distributes the spark more evenly than the pin-cathode arrangement. Recirculation fans move cool gas through the lasing region, increasing the population inversion.

The second type of high-power CO_2 laser, originally developed for military applications, is the gas dynamic laser. In the gas dynamic laser, a population inversion is obtained through the application of thermodynamic principles rather than by standard discharge-tube techniques. A nitrogen-carbon dioxide gas mixture is first heated, then compressed, and then sent through a specially designed nozzle into a region of greatly reduced pressure. During the heating-compression cycle, the population of energy states reaches the Boltzmann distribution appropriate to the higher temperature. At low temperatures, most of the energy stored by the diatomic nitrogen molecule is stored as translational and rotational energy. At higher temperatures, however, a sizable fraction of the energy is stored in the vibrational mode of the N_2 molecule, a stretching of the interatomic bond. While translational and rotational energy is rapidly lost to the surroundings through collisions when the temperature is suddenly reduced, the vibrational energy is not so easily dissipated. The nitrogen molecular vibrational levels act as an energy storage medium. It is the resonant collision of the excited N_2 molecules with the CO_2 that populates the (001) state of CO_2 and creates the inversion. The population inversion can be increased through the addition of small amounts of water vapor at the points of expansion of the gas. This additive hastens the relaxation of the (100) state to the ground state. An optical resonator is located within the region where the population inversion is maximum. With very active pumping, CW output powers of many tens of kilowatts can be obtained from such devices. A shock-tube-driven laser has achieved a multimode output power of 400 kW for a duration of 4 msec. The disadvantages of the gas dynamic laser are its bulk and its rocket-like roar that accompanies the rapid gas expansion. In fact, many of the elements of rocket technology are being applied to the development of this laser.

Chemical Lasers

The lasers studied in this chapter are classified primarily on the basis of the state of the active medium (gas, solid, liquid, semiconductor). The term *chemical laser* refers not to the state of the lasing medium, but to the method of creating a population inversion. In the chemical laser, the excitation is produced by a chemical reaction. Although the chemicals can be in the solid, liquid, or gaseous state, most chemical lasers use gases as the active medium, with an arrangement similar to the gas dynamic CO_2 laser just discussed.

Chemical lasers are attractive from several viewpoints. A purely chemical laser, relying on the direct mixing of chemicals to produce coherent light, does not require electronic components or electrical power. Chemical lasers have the potential for higher output power per unit volume and per unit weight than appears possible with electrical excitation. Because the chemical reactions employed excite primarily vibrational states rather than electronic states, most chemical lasers have output powers in the near infrared, with wavelengths between the neodymium laser at $\lambda = 1.06 \ \mu m$ and the carbon dioxide laser at $\lambda = 10.6 \ \mu m$. Chemical lasers have produced some of the most powerful laser pulses ever observed. Pulses as large as 4200 joules with a peak power of 200 billion (2×10^{11}) watts have been achieved by a hydrogen fluoride chemical laser.

Most of the chemical reactions used in chemical lasers are of the form

$$A + BC \rightarrow AB + C + \text{energy} \qquad (6.1)$$

The energy released by the reaction serves to excite the molecule AB, which serves as the active medium. One reaction that has been investigated extensively is the reaction of a halogen with hydrogen or deuterium, e.g.,

$$H + Cl_2 \rightarrow HCL + Cl + \text{energy} \qquad (6.2)$$

and

$$F + D_2 \rightarrow DF + D + \text{energy} \qquad (6.3)$$

In general, chemical reactions can be employed successfully in laser systems using several approaches:

1. Chemical reactions can directly produce the radiant energy from the reacting species, with the addition of no external energy.

2. Chemical reactions can result in light emission from the reacting species, but external energy may be necessary to initiate or sustain the reaction.

3. External energy can be provided to initiate or sustain a chemical reaction that results in the transfer of energy from the reacting species to another species that radiates.

4. The chemical reaction can provide all the energy, as in type (1), but energy is transferred from one reacting species to another species that emits radiation.

Lasers using the latter two approaches, in which the reacting species do not actually participate in the lasing action, are called *chemical transfer lasers*. The reactions represented by Eqs. (6.2) and (6.3) are examples of approach (2). Although the chemical reactions produce HCl or DF (the active lasing medium) in the excited state, the dissociation of the hydrogen or fluorine atoms from their initial molecular states (H_2 or F_2) must be accomplished with an additional energy source, a flashlamp in the case of reaction (6.2), or a thermal source in the case of reaction (6.3).

Another example of approach (2) is the carbon monoxide chemical laser, which emits relatively high power in the infrared. A mixture of helium, air, and cyanogen (C_2N_2) has been used to obtain several watts of power in the wavelength range between 5 and 6 μm. In this laser, the gas is flowed through an electric discharge. The helium does not participate directly in the lasing, but provides resonant transfer of energy, as in the helium-neon laser. The discharge dissociates the C_2N_2 molecules and the O_2 molecules to form vibrationally excited CO via the reaction

$$C_2N_2 + O_2 \rightarrow 2CO + N_2 + 127\,kcal \quad . \tag{6.4}$$

Lasing occurs between the vibrational levels of the CO molecule.

The construction of chemical lasers varies considerably with the type of reactions and chemicals. Many chemical lasers, such as the deuterium fluoride (DF) laser shown schematically in Fig. 6.11, closely parallels the construction of the gas dynamic laser. An inert carrier gas such as helium or nitrogen flows through the entire structure. One reactant, usually the fluorine, is added directly to the carrier gas. The deuterium is injected through a multiple nozzle assembly, and the reaction occurs between the mirrors of the optical cavity. The beam is emitted perpendicular to the gas flow.

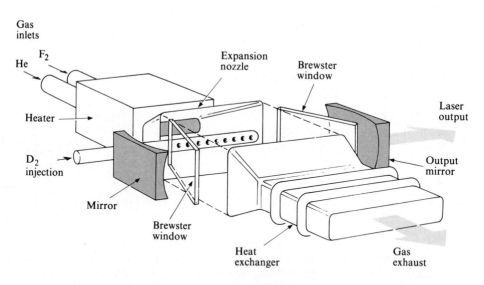

Fig. 6.11 Schematic of a chemical laser. One of the chemical reactants (in this case, F_2) is heated with a carrier gas (He) and allowed to expand just before mixing with the second reactant (D_2). The reaction takes place in the region between the two Brewster windows. (The enclosure around this area has been omitted for the sake of clarity.) The output beam is in a direction transverse to the gas flow, as in the gas dynamic laser.

Other Molecular Lasers: Nitrogen and Excimer

Since the number of molecular systems that have energy-level transitions suitable for lasing action is enormous, it is probable that many additional sources of laser light will be discovered. Two molecular lasers that emit in the ultraviolet portion of the spectrum have been developed: the nitrogen laser and the excimer laser.

The properties of the nitrogen (N_2) laser are quite different from those of the CO_2 laser. In the CO_2 laser, the transitions correspond to vibrations of the CO_2 molecule and produce radiation in the infrared; in the N_2 laser, the transitions correspond to different electronic energy states, and the emitted radiation is in the ultraviolet at 337.1 nm. This relatively high-frequency output, produced in short, high-power bursts, makes the nitrogen laser an ideal pump for other laser systems.*

A prerequisite for continuous-wave laser operation is that the upper state of the lasing transition must be long-lived and the lower state be rapidly depopulated. In the nitrogen laser, the converse is true. The upper-state lifetime is very short—on the order of 5 nsec—and the lower-state lifetime is three orders of magnitude longer. As a consequence, the population inversion, and therefore lasing action, cannot be maintained for a duration much longer than the lifetime of the upper state. As the number of atoms in the lower state increases, the stimulated output is reabsorbed by these atoms, causing a large loss that increases with time and finally shuts down laser action. This type of laser system is known as *self-terminating*. Very fast risetime pumping mechanisms must be used to create a population inversion.

The configuration of many nitrogen lasers resembles that of CO_2 TEA lasers. A large capacitor is discharged through nitrogen gas contained between two long, parallel electrodes. Lasing action is obtained parallel to the electrodes, perpendicular to the electric discharge path. Several novel switching methods for the high voltage have been developed. For cooling, the gas is flowed along the lasing path. The nitrogen laser differs from other gas lasers that we have discussed in that the gain produced by the population inversion is so large that the requirement for feedback is not applicable in most cases. As was pointed out in Section 3.3, removing the optical resonator changes the laser from an oscillator to an amplifier, yet it is still termed a laser. In some nitrogen lasers, a single mirror is used to effectively double the amplification path by redirecting amplified light traveling in the direction of the mirror back through the amplifying medium to the output end of the laser. Such high-gain operation, where no cavity feedback is used, is termed *superradiant*. The single mirror effectively doubles the pathlength of the

---

* We note from our discussions of pumping mechanisms that lasing action cannot occur at a frequency higher than the pump frequency unless nonlinear pump processes, such as a two-photon absorption process, are involved. Generally, such nonlinear mechanisms require high input power and are not efficient for laser pumping.

laser medium. Because of the elongated transverse discharge region, the divergence of the output beam is much greater in the plane perpendicular to the discharge, than it is in the plane of the discharge. The output beam is therefore fan-shaped, similar to the semiconductor diode laser discussed in Section 6.3.

Commercial pulsed nitrogen laser systems are capable of 100 kW of peak power in 10-nsec pulses, with a 500-Hz repetition rate. Because they generate intense ultraviolet radiation in short bursts, nitrogen lasers are particularly suited to photochemical investigations, because the photon energy is sufficient to initiate reactions by molecular dissociation or by excitation to states that cannot be reached with visible lasers.

The excited states of the nitrogen molecule are so short-lived that they cannot serve as an efficient energy storage medium. Yet for practical ultraviolet and visible lasers and amplifiers, it is this requirement of an efficient storage medium that must be met. This storage may be achieved with an active medium that is a bound-free system. The constituents of such a system, when in their ground state, repel one another at interatomic distances characteristic of most diatomic molecules. They are, in effect, "free" of one another. For some atoms, an excitation or ionization sufficiently modifies the state of the atom so that there is an attractive force with other atoms in the gas. The two atoms are then bound together at a small separation distance, creating an *exc*ited-state d*imer*, or *excimer*. If one or both of the excited-state atoms in the system are rare-gas atoms, the amount of excitation energy is extremely large; thus the metastable excimer state is an important system for storing high energies.

The active medium in an excimer laser can be an excited rare-gas dimer (Ar_2^*, Kr_2^*, Xe_2^*; * denotes excited state), a rare-gas oxide (ArO^*, KrO^*, XeO^*), or a rare-gas atom in combination with a halide atom (ArF^*, KrF^*, XeF^*). In some cases, the excimer state is not the upper state of the laser transition, but serves as a pump state, just as the excited helium atoms do in the HeNe and HeCd lasers. For example, the Ar_2^* state can pump molecular nitrogen (N_2) to an excited state where it emits in the near ultraviolet or blue-green region of the visible spectrum. In the pure excimer systems, lasing has been obtained in the ultraviolet at 125 nm (Ar_2^*), 146 nm (Kr_2^*), and 172 nm (Xe_2^*). All of the rare-gas oxides have lased in the vicinity of 558 nm, since the lasing transition for all of these systems occurs between two excited states of atomic oxygen. The upper state is long-lived and not easily deactivated by collisions with other atoms, whereas the lower state is rapidly depopulated by collisions. This is an ideal set of circumstances. The transition at 558 nm is responsible for the yellow-green glow of the polar sky displays known as auroras. Thus the laser using this transition might be rightly termed an "aurora" laser.

These systems have been pumped by an intense electron-beam source or by a fast discharge similar to the N_2 laser. Electron beam pumping is achieved with a series of high voltage potentials that accelerate electrons to energies of 1 MeV. The pulse of electrons is formed and transmitted to the laser chamber at beam currents on the order of 100 kiloamperes. The beam is transmitted through a 130-μm thick titanium foil, which separates the low-pressure electron generator space from the high-pressure (as high as 60 atm) reaction chamber.

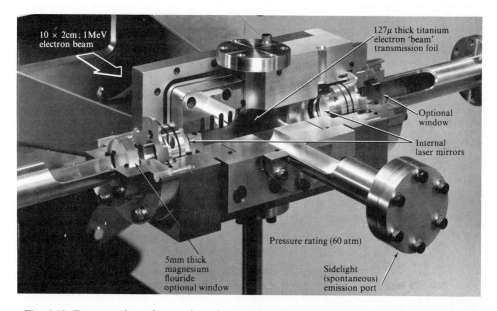

Fig. 6.12 Cutaway view of an excimer laser cavity. The excitation is in the form of a high-current electron beam passing through the titanium foil. The laser mirrors are internal to the cavity. The 60-atm pressure is contained by two windows external to the laser cavity. (Courtesy of Lawrence Livermore Radiation Laboratory, University of California.)

A laser cavity for an electron beam-pumped excimer laser is shown in a cutaway view in Fig. 6.12. The active volume is about 2.5 cm in diameter and 10 cm long. The mirrors of aluminum on a magnesium fluoride substrate are separated by about 15 cm to form the laser cavity. Separate magnesium fluoride windows outside the cavity serve to contain the high pressure. In addition to the laser light output windows, there is an additional side port to monitor spontaneous emission of the discharge for diagnostic purposes. Besides a change in the frequency distribution of the light output and the appearance of a collimated beam when lasing begins, there is also a drop in the spontaneous emission output at right angles to the laser beam.

Research on excimer lasers is being pursued vigorously, not only because of the ultraviolet wavelengths with coherent output, but also because of their possible use in laser fusion research. The overall efficiencies of some of the excimer lasers appears promising. A number of the rare-gas halide lasers have been operated with pumping efficiencies of 10 to 15 percent. A further discussion of the applications of excimer lasers is found in Section 9.3.

6.2 DOPED-INSULATOR LASERS

Doped-insulator lasers occupy a unique place in laser development; the first operational laser medium was pink ruby. The term "doped-insulator laser"* is used to describe a laser whose active medium is a regular array of atoms: a crystal with impurity ions intentionally introduced into the crystal at the time of growth (a process called *doping*). These lasers are rugged, simple to maintain, and capable of generating high peak powers. For these reasons, their applications are diverse, ranging from resistor trimming to the proposed initiation of thermonuclear fusion.

The Active Medium

In most cases, the active medium of a doped-insulator laser consists of an impurity or *dopant* in a crystalline insulator. The crystal atoms do not participate directly in the lasing action, but serve as a host lattice in which the dopant resides. The dopant usually substitutes directly into the host crystal lattice, rather than being an interstitial impurity (an impurity residing between lattice points). The dopants can be considered as a "frozen gas" of heavy ions randomly distributed throughout the crystal. When an ion is free to move in a gaseous discharge, rather than constrained in the lattice, many of the energy levels have the same energy and are said to be *degenerate*. When the ion is placed within the lattice of a crystal, however, the crystalline electric field causes some electron distributions to be more favorable than others, and the energies are no longer equal. The crystalline field partially removes the degeneracy.

To better understand this degeneracy, consider the following analogy. Suppose you were in free fall in a starless, planetless universe. You would have no preference for which direction you faced. All directions would be the same; they would be degenerate from your point of view. Now, suppose that eight identical spaceships appeared in formation about you, inspecting this piece of cosmic debris. Each spaceship marks a corner of a cube with you at the center of the cube, as shown in Fig. 6.13. Now you have some preferences. You can look toward a spaceship, you can look at a line of the cube between two spaceships, or you can look at

---

* The term "solid-state laser" is also used for these lasers. However, because the semiconductor laser is a solid-state device, we prefer to label the lasers under discussion as doped-insulator lasers. This term should reduce confusion and serve as a more descriptive label.

a face of the cube, i.e., at a point equidistant from four spaceships. Certain directions are now different from other directions. The appearance of the saucers has removed some (but not all) of the degeneracy of direction in that featureless space. This is analogous to the removal of degeneracy by a crystalline field.

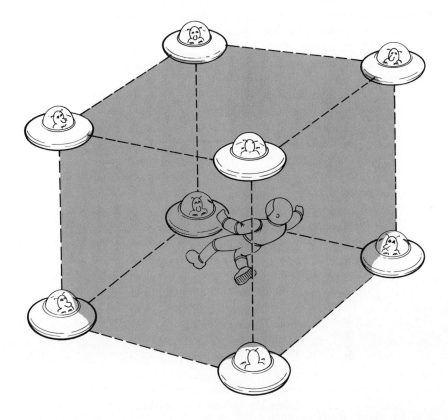

Fig. 6.13 An analogy to crystal field splitting of the previously degenerate levels of an ion in a crystal (see text).

The crystalline host is very important in determining the absorption and emission characteristics of dopants. Consider the case of the triply-ionized chromium atom, Cr^{3+}, which is the dopant for the ruby laser. In the free state, the lowest energy level of the chromium ion corresponds to 28 degenerate quantum states. When bound in the host lattice of an aluminum oxide crystal, the chromium ions absorb strongly in the blue and green portions of the visible spectrum.* When in the free state, however, as would be the case in a gas discharge, no such absorption is observed. The underlying reason for this difference in absorption (and emission) characteristics is the splitting of a number of energy states of the ion when it is placed in the crystal lattice. The energy-level diagrams for the two cases are illustrated in Fig. 6.14. As noted in the figure, most of the energy levels fall into one of two bands of levels, labeled ② and ③ in the figure. Transitions from the lowest energy level, labeled ①, to levels in groups ② and ③ correspond to the observed absorption in the green and blue portions of the spectrum. The linewidths of these transitions are substantially broadened by collision (thermal) and other broadening mechanisms in the crystalline lattice, and the groups of levels are, as a consequence, usually referred to as *absorption bands*.

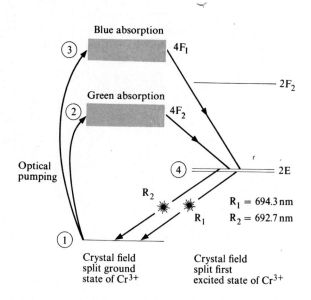

Fig. 6.14 Energy-level diagram for the ruby laser. The wavelengths of the two laser lines, R_1 and R_2, depend upon temperature. The values given here are typical.

When the ion is placed in the crystalline field, the first excited state also splits into a number of energy states, several of which, labeled ④, lie just below the

---

* The beautiful deep-red coloration of ruby is due to chromium ion dopants, which absorb light in the blue-green portion of the spectrum.

lowest absorption band, labeled ②. The ruby laser operates on the basis of absorption transitions from the lattice ion ground state, the lowest level shown in Fig. 6.14, to the green and blue absorption bands. The ions then undergo nonradiative transitions to two lower-lying metastable energy states ④. These states are the upper levels of the two ruby laser transitions. The two corresponding emission lines, called the R_1 and R_2 lines, are at 694.3 nm and 692.7 nm, respectively. Stimulated emission at the 694.3-nm line usually dominates. Note that the ruby laser is based on a three-level pumping scheme in that the ground state is the lower level of the laser transition.* As discussed earlier, such lasers have a high threshold. Only by vigorous pumping can a population inversion be attained between these metastable states and the ground state.

A second, very important doped-insulator laser is the neodymium laser. The active medium for this laser is the rare earth Nd^{3+} ion doped into host insulator materials of glass or yttrium aluminum garnet (YAG). As with the ruby laser, multiple low-lying energy levels result when the ion is placed in the host material. Although glass is an amorphous solid rather than a crystal, the local electric fields within the material also split the initial energy states of the ion. Unlike the case for the crystal-doped ion, the field at an ion may differ from that at others. For this reason, the split energy levels are less well-defined, and the linewidth of the Nd:glass laser is greater than that of the Nd:YAG laser. Neodymium has an output in the infrared portion of the spectrum, the predominant wavelength being at 1.06 μm. Because the neodymium lasers are four-level systems, their pumping thresholds are considerably smaller than that of the ruby laser.

Pumping

To excite the laser ions to the higher electronic states (② and ③), a flashlamp is generally used. Historically, the first pumping lamp was a xenon flashlamp used for photography. A ruby rod was inserted into the center of the helical flashtube and the lamp was fired. The polished and silver-coated ends of the ruby rod served as the resonator mirrors. Since that time, flashlamps and their associated electronics have become quite sophisticated, but the basic method remains unchanged. Since the absorption bands extend from the visible into the ultraviolet, it is efficient to use flashlamps that radiate throughout this region. As a consequence, most flashlamps have quartz envelopes instead of glass, which absorbs ultraviolet radiation. A photograph of the interior of the laser head of a commercial ruby laser is shown in Fig. 6.15. This laser uses a helical flashlamp, liquid coolant, and external mirrors.

---

* It should be noted that the above scheme is a simplification of the energy splittings in these crystals. In some cases, the ground-state energy levels can be split again by applying a magnetic field. These additional energy separations are such that the system can be used as a *maser*, to emit coherent microwave radiation rather than light.

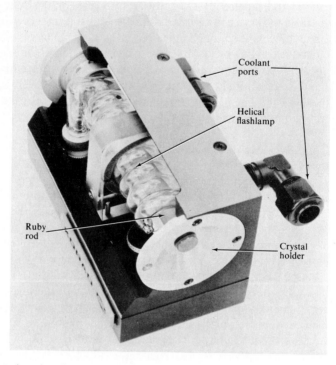

Fig. 6.15 Interior of a ruby laser head. (Courtesy of Korad, Division of Hadron Corporation.)

Besides the helical lamps, cylindrical xenon flashlamps are also employed for pumping doped-insulator lasers. In contrast to the helical lamps, the cylindrical lamps are easy to cool, since they require only an extra outer cylinder over the flashtube to provide space for flowing coolant. The geometry of the pumping cavity varies considerably with the particular application. In many of these lasers, one or two cylindrical lamps are placed parallel and adjacent to the laser rod. Another arrangement takes advantage of the focusing properties of an ellipse. The laser rod and flashlamp are placed inside a cylindrical reflector of elliptical cross section, with the rod at one focus and the lamp at the other focus. Any light leaving one focus of the ellipse will pass through the other focus after reflection from the silvered surface. In some cases, two or four ellipses are used, with the rod at the common focus.

The flashlamps in a doped-insulator laser cavity are usually operated at current densities greater than 100 amperes/cm^2. The lamps are connected in series with an energy storage bank consisting of one or more capacitors and with an inductor that limits the peak current through the lamps. The inductor also shapes and lengthens the current pulse, protecting the lamps from damage, extending their useful life, and giving longer-duration output pulses. A schematic diagram of a flashlamp driving circuit is shown in Fig. 6.16. The flashlamps are fired by energizing the bank to a voltage that is less than the breakdown voltage of the flashlamps, and then applying a trigger pulse to the primary of a transformer. The secondary of the trigger transformer is in series with the lamps. The high-voltage pulse (20 kV) ionizes the gas inside the flashlamp tube and provides a high-current conducting path to discharge the energy bank.

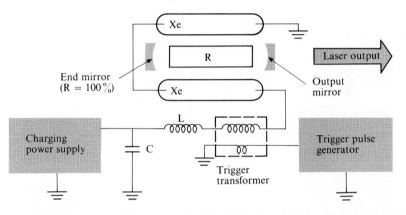

Fig. 6.16 Schematic of a simple flashlamp-pumped laser. The trigger pulse generator and transformer provide a high-voltage pulse sufficient to cause the xenon gas in the lamps (Xe) to discharge. The ionized gas provides a low-resistance discharge path to the storage capacitor (C). The inductor (L) shapes the current pulse, maintaining the discharge. The discharge of the lamps optically pumps the laser rod (R).

Normally a countdown procedure is employed before each firing of a doped-insulator laser system, and users shield their eyes. From the standpoint of safety, the flashlamp power supplies are probably as dangerous as the laser output itself. Although the supply may not appear to be a threat, a large capacitor bank of several millifarad capacitance charged to a potential of several thousand volts can deliver a lethal jolt. For this reason, a well-designed power supply contains interlock mechanisms that automatically discharge the capacitor banks when the power supply is opened for inspection. A well-grounded probe with an insulating handle is frequently attached to the door of the power supply to drain off any residual charge on the capacitor bank.

Laser Cavity

As with the first ruby laser, the crystal itself may serve as the optical resonator. The ends of the crystal, which is usually in the form of a rod, are lapped, polished, and coated for proper reflectivity at the laser wavelength. Since mirror coatings are subject to damage, roof prisms or corner cubes are sometimes fashioned on one end so that the light is reflected by total internal reflection. Most doped-insulator lasers today use external mirrors. In this case, it is desirable to prevent reflections between the faces of the rod and the external mirrors. For low powers, special antireflection coatings can be deposited on the laser rod, or the rod can be tilted at an angle to the axis of the laser cavity. For high-power applications, the laser rod is often cut at the Brewster angle and mounted as shown in Fig. 6.17. This configuration offers several advantages. It allows external devices, such as Q-switching apparatus, to be placed inside the laser resonator. Since the rod is cut at the Brewster angle, no additional boundary conditions are introduced that would tend to suppress oscillation of a large number of transverse modes. The presence of many modes is necessary for the generation of short-duration, high-power laser pulses, as we noted in the previous chapter. The Brewster angle cut also guarantees a polarized output.

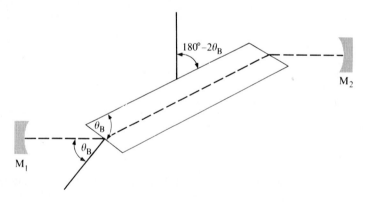

Fig. 6.17 External-mirror, doped-insulator laser with the laser rod oriented at the Brewster angle.

Although doped-insulator lasers offer some unique advantages over gas lasers, crystals are not ideal cavities or perfect laser mediums. Real crystals contain refractive-index variations that distort the wavefront and mode structure of the laser. High-power operation causes thermal expansion of the crystal that alters the effective cavity dimensions and thus changes the modes. The laser crystals are cooled by forced air or forced liquid, particularly for high repetition rates. In some cases, liquid nitrogen is used to cool the laser rod. For very high-power doped-insulator lasers, the laser rod itself may be segmented, with liquid of the same refractive index as the rod flowed between the segments. This effectively increases the surface area of the active medium and enhances cooling. A segmented-type construction which uses neodymium-doped glass discs is discussed in Section 9.1 on high-power lasers.

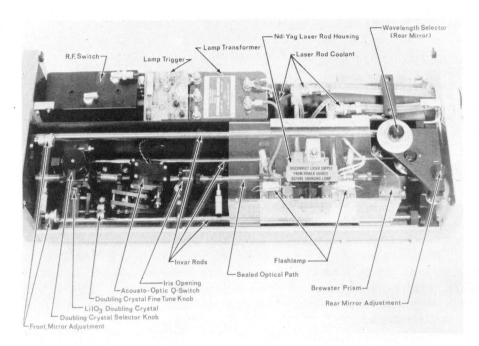

Fig. 6.18 Commercial Nd:YAG laser including frequency-doubling and Q-switching apparatus. Laser output at 1.06 μm is converted to 0.53 μm by frequency-doubling. (Courtesy of Chromatix.)

A detailed photograph of a commercial Nd:YAG laser system is shown in Fig. 6.18. This laser has a frequency-doubling crystal inserted between the Nd:YAG rod and the output mirror. The frequency-doubling crystal, discussed again in Section 6.5, converts the 13 infrared output wavelengths to 13 visible wavelengths between 473 nm and 679 nm. The infrared wavelengths are selected with a Brewster prism arrangement similar to that described in Section 5.1. For Q-switching, an acoustooptic shutter is used.

Output

The most striking aspect of doped-insulator lasers is that the output is usually not continuous, but consists of a large number of bursts of power, or spikes, as discussed in Section 4.3 on gain and in Section 5.3 on Q-switching. Because of this spiked output, the coherence length of these lasers is frequently only several centimeters, quite short compared to gas lasers. Although this short coherence length makes doped-insulator lasers unsuitable for many applications, they are unsurpassed for producing intense bursts of visible light from a relatively small laser. The Q-switching and mode-locking techniques discussed in the previous chapter are used with such lasers.

6.3 SEMICONDUCTOR LASERS

Semiconductor lasers are the cheapest and smallest lasers available. They are commercially significant because they can be mass-produced and easily fabricated into arrays using the same techniques developed for transistors. Because of their efficiency and small size, they are well suited as light sources for fiberoptic communication systems.

Electronic Properties of Solids

To understand the operation of a semiconductor laser, we must first discuss the properties of electrons and their distribution within the possible electronic energy states of a semiconductor. All electrons, whether tightly bound in an atom or mobile as in a metal, obey the Pauli exclusion principle, which states that only one electron is allowed in each possible quantum state (see Section 3.1). This means, first, that only two electrons can occupy the lowest possible energy state (the two have opposite spins and therefore do not violate the exclusion principle), and second, that an electron cannot be excited to a higher state unless the higher state is unoccupied.

As a consequence of the exclusion principle, a large number of electrons in a solid are in energy levels considerably above the lowest energy level. At a temperature of absolute zero, all the electronic energy levels up to a given level are occupied; all higher energy levels are unoccupied. The energy level that divides the occupied from the unoccupied levels is called the *Fermi energy*. As the temperature is raised above absolute zero, thermal excitation of electrons populates some of the unoccupied higher states. The Fermi energy is no longer a dividing line between occupied and unoccupied states, but is instead the energy at which the probability of occupation of a state is one-half (there is a fifty-fifty chance of an electron being in that state). We may still think of the Fermi energy, however, as the energy level above which very few states are occupied.

The electronic properties of a solid are determined largely by the periodic crystal lattice arrangement of the atoms that make up the solid itself. A quantum-mechanical calculation of the allowed energy levels for electrons in a periodic lattice predicts that the allowed electronic states occur within well-defined bands of energy. Between these bands are energy gaps, ranges of energies that no electron in the solid can possess. The highest filled band in the electronic energy distribution is the *valence band*. Those states that are either partially or completely unoccupied by electrons form the *conduction bands* of the solid. It is the size of the energy gap E_g (referred to as the "band gap") between the valence band and the lowest conduction band that determines whether a solid is an insulator, a metal, or a semiconductor. If the valence band is completely occupied and the conduction band is empty, as shown in Fig. 6.19(a), the material is an insulator. This is because conduction is due to the movement of "free" charge carriers from one state in

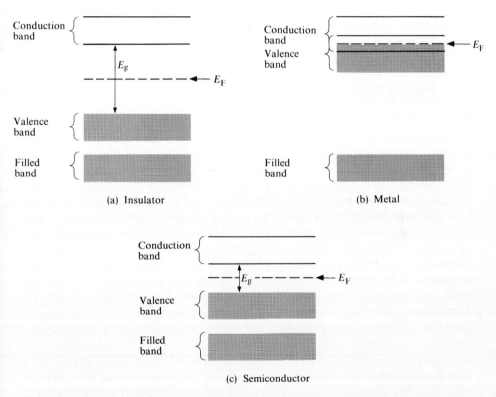

Fig. 6.19 Energy-band diagrams of (a) an insulator, where the band gap is large compared to thermal excitation energies ($E_g \gg kT$); (b) a metal, in which the valence and conduction bands overlap (as shown here), or the conduction band is only partially filled; (c) a semiconductor, where the band gap is of the same size as the thermal excitation energy ($E_g \doteq kT$). E_F indicates the position of the Fermi energy (see text).

a band to another state in a band under the influence of an electric field. This cannot occur in a completely filled band, since all possible states that an electron could occupy during conduction are already occupied. In essence, there is "no place for the electron to go." The nearest available states in energy are in the lowest conduction band, and the gap between the valence band and this conduction band is so large that thermal excitation is insufficient to excite electrons into the conduction band ($E_g \gg kT$). In contrast, metals are good conductors because either there is an overlap between the valence and conduction bands, or the conduction band is partially filled (Fig. 6.19b). Under the influence of an electric field, the electrons are free to move because there are nearby empty states available.

If the band gap for a material is sufficiently small that thermal excitation can excite electrons in the valence band to states in the conduction band ($E_g \simeq kT$), the conductivity is intermediate between that of a metal and that of an insulator. The material behaves as a semiconductor (Fig. 6.19c). When an electron in the valence band of a semiconductor is excited thermally to the conduction band, an unfilled energy state in the valence band occurs. This unfilled state, called a *hole*, has many of the same properties as an electron, except that a positive charge is assigned to it. In the presence of an electric field, the hole moves in the direction of the field, just as a positive charge would.

Semiconductors

A semiconductor with no impurities, called an *intrinsic semiconductor*, has an equal number of conduction electrons and holes, since for every conduction electron produced there must be a corresponding hole generated. The number of conduction electrons and valence-band holes can be increased by heating the semiconductor, because, by the Boltzmann law (Section 3.2), the states higher in energy have a greater probability of being occupied as the temperature is increased. The number of conduction electrons or holes can be altered by adding impurities,—i.e., by doping the semiconductor. A crystal composed of a collection of atoms with four valence electrons, e.g., silicon, shares each of its electrons with each of the four surrounding atoms by covalent bonding; all four valence electrons of each atom are tightly bound into a stable lattice structure. If this crystal is doped with atoms having five valence electrons, the extra electron, with no covalent bond to complete, will have a much smaller binding energy than its eight neighbors. This extra electron is then easily excited into the conduction band and behaves like a free electron within the crystal. This situation is shown in Fig. 6.20(a). Such a doped semiconductor material is referred to as *n-type* (n for negative), and the dopant atoms are called donors. If we dope the crystal with atoms having three valence electrons, as shown in Fig. 6.20(b), we find that the vacancy created by the incomplete covalent bond is quite mobile and acts as a hole. In this case the semiconductor is referred to as *p-type* (p for positive), and the dopant atoms are called *acceptors*.

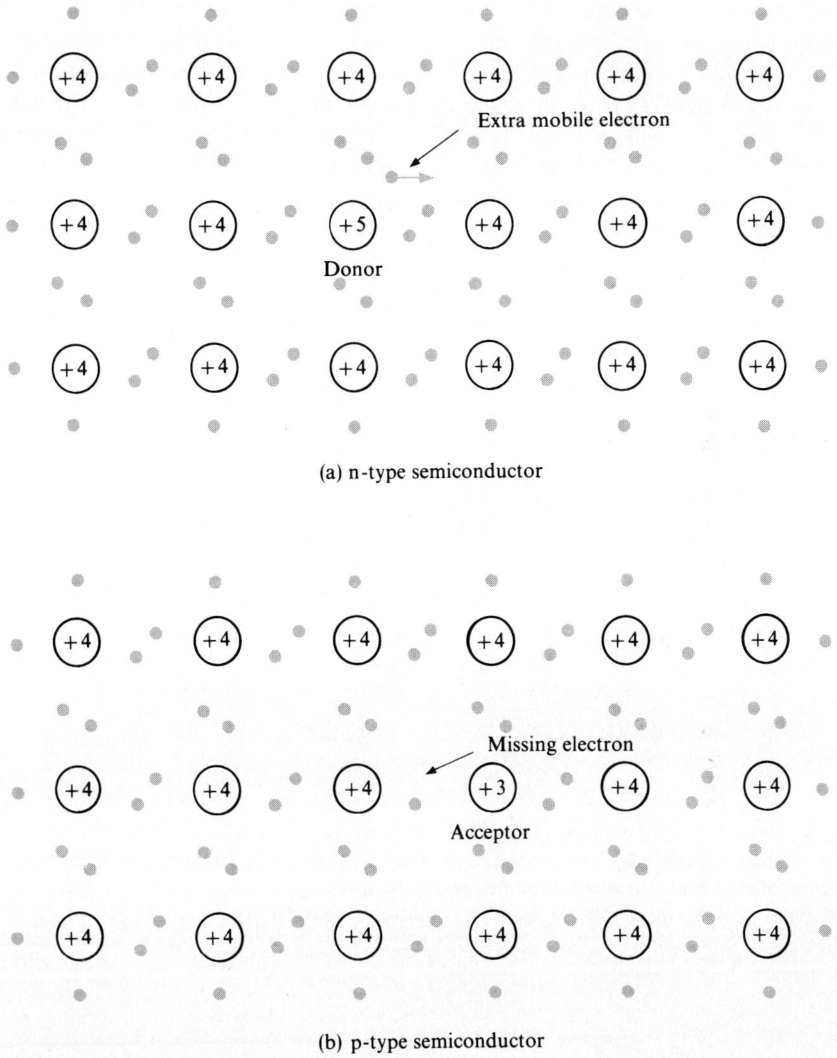

(a) n-type semiconductor

(b) p-type semiconductor

Fig. 6.20 (a) An n-type semiconductor, showing one donor atom. (b) A p-type semiconductor, showing one acceptor atom.

Radiation From a p–n Junction

The most common form of semiconductor laser, the semiconductor diode laser, consists of a p-type and an n-type material joined together. The energy-level diagrams for p-type and n-type material are shown in Fig. 6.21. In the n-type, we show a group of donor energy levels (energy E_D) just below the conduction band, and in the p-type a group of acceptor levels (energy E_A) just above the valence band. For moderate doping on the n-side, a portion of the donor levels are occupied and the Fermi energy E_F lies between the conduction band E_C and the donor level E_D. For moderate doping on the p-side, a portion of the acceptor levels are unoccupied and the Fermi energy lies between the acceptor level E_A and the valence band E_V.

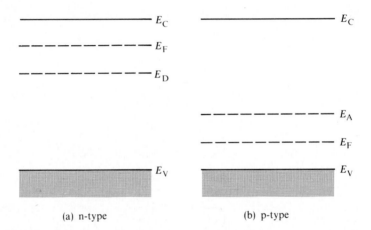

(a) n-type   (b) p-type

Fig. 6.21 Energy-band diagrams for n-type and p-type semiconductors. E_V and E_C correspond to the energies of the valence and conduction bands, respectively. In the n-type semiconductor, additional electrons are weakly bound, giving them energy levels E_D just below the conduction band. The Fermi energy E_F is to be found in the vicinity of E_D. Similarly, in the p-type semiconductor, creation of electron vacancies (holes) gives rise to energy levels E_A just above the valence band, with the Fermi level just below it.

When we bring the two types of materials into contact, forming a *p–n junction*, thermodynamics requires that the energies of the various levels shift relative to one another until the Fermi energy E_F is continuous throughout the medium, as shown in Fig. 6.22. The energies of the valence and conduction bands differ on the two sides of the junction region, with the p-side being at the higher potential energy. If we apply a positive voltage to the n-type side of the p–n junction and a negative voltage to the p-type side, the two carriers flow toward the electrodes. Once all the electrons on the n-side enter the positive electrode, no more free electrons are available, and the current flow ceases. This is referred to as *reverse biasing* the junction. If the voltages are reversed, with the positive voltage on the p-type region (a condition referred to as *forward bias*), the electrons attracted to the positive electrode cross into the junction region. There they will recombine with the holes, which are attracted to the negative voltage. The battery resupplies the n-type region with electrons and creates more holes in the p-type region by depleting it of electrons. The current continues to flow. Thus the p–n junction acts as a diode, conducting current when a positive voltage is applied to the p-side, not conducting when the voltage is reversed.

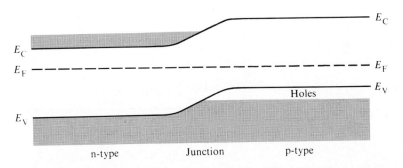

Fig. 6.22 Energy-band diagram for a p–n junction with moderate doping. Shaded regions represent electron concentrations.

A semiconductor laser requires that there be a region of the p–n junction where both electrons and holes are present. This can be obtained by heavily doping both sides of the junction. With heavy doping on the n-side, both the donor levels and a portion of the conduction band are occupied; hence the Fermi energy E_F lies within the conduction band. With heavy doping on the p-side, the acceptor levels are unoccupied, and holes exist in the valence band; the Fermi energy lies within the valence band. When the two materials are brought into contact, the levels shift, and the Fermi level is again uniform throughout the material. The resultant energy diagram is shown in Fig. 6.23(a). When a forward bias is applied to the diode, the energy levels shift to a new distribution, as shown in Fig. 6.23(b). The important thing to note is that the junction is a narrow region where both electrons and holes exist.

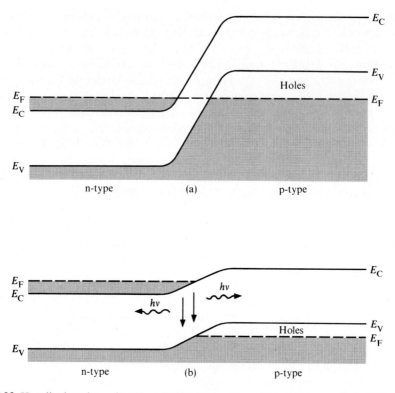

Fig. 6.23 Heavily doped p–n junction. (a) Junction with no bias voltage applied. (b) Forward biased ($+ V_B$ applied to the p-side). Note that the Fermi levels are now inside the valence and conduction bands due to the heavy doping. The difference in energy between Fermi levels is equal to eV_B (e is the electronic charge).

Optical radiation can occur in this narrow region when the diode is forward biased (conducting). When an electron in the conduction band recombines with a hole in the valence band, a quantum of radiation with energy equal to the energy difference between the two states is released. For many semiconductors, this energy difference corresponds to a photon of visible or near-infrared radiation. This radiation is called *recombination radiation*. If we can achieve a sufficient population inversion (a greater number of electrons in the conduction band than holes in the valence band) in the junction region, then the recombination may be stimulated and laser action is possible. When the emission is not stimulated, such devices are called *light-emitting diodes* (LED). They serve as highly efficient, low-voltage indicators or display lamps replacing the gas discharge tubes and other vulnerable high-voltage display components. Most commercially available LED outputs are confined to the red and orange portions of the spectrum. For most semiconductor lasers the output radiation is in the infrared.

The wavelength of the transition is determined by the size of the band gap. The potential energy needed to raise an electron from the top of the valence band to the bottom of the conduction band must be equal to the energy of the photon emitted upon recombination. If V_g is the electrical potential needed to promote the electron to the conduction band and e is the electronic charge, the corresponding potential energy is eV_g. We therefore have

$$E = hv = eV_g.$$ (6.5)

The wavelength of recombination radiation is then

$$\lambda_{recomb} = \frac{c}{v} = \frac{hc}{eV_g}.$$ (6.6)

If h is expressed as 4.14×10^{-15} electron-volt-sec and e is one electronic charge, the gap voltage for a laser with a 890-nm output wavelength is 1.4 volts.

Semiconductor Lasers

To make a light-emitting diode lase, we must meet the necessary requirements for lasing action. The medium for laser action is already present in the form of the junction region, so we merely have to increase the current until a population inversion is reached and add some mirrors to provide feedback. One of the simplifications of diode lasers compared to gas lasers is that no external mirrors are necessary. The high reflectivity due to the refractive-index difference at the diode-air interface is sufficient. The diodes are cleaved along natural crystalline planes. The parallelism of reflecting surfaces is thus assured and further polishing of the optical surfaces is unnecessary.

Usually, room-temperature semiconductor lasers are pulsed. To take advantage of the fast risetime that is possible with semiconductor lasers, specially designed pulse circuits must be employed. These circuits are similar to the solid-state flashlamp circuit illustrated in Fig. 6.16, since the energy to pulse the laser is produced by discharging a capacitor. Between pulses, the power supply charges the capacitor through a charging circuit, which may consist of a single resistor. During the charging cycle, the capacitor is prevented from discharging by the silicon controlled rectifier (SCR), a three-terminal, solid-state switch that conducts when a trigger pulse is applied to its third terminal (called the *gate*). Once the SCR has been turned on, it continues to conduct until the capacitor is discharged. The peak current obtained with such a circuit may be on the order of 50 amperes, with a pulsewidth of 100 nanoseconds, producing a peak laser power of approximately 10 watts.

The onset of lasing of a light-emitting diode is not as dramatic as that of a typical gas laser. As the current of an LED is increased to threshold, the output irradiance is seen to increase abruptly in the direction of the laser beam (Fig. 6.24). The beam divergence angle decreases, and the spectral width of the emitted light narrows greatly, as illustrated in Fig. 6.25. The threshold current for laser operation is highly dependent on the temperature of the diode. Between room temperature (~ 300 K) and the temperature of liquid nitrogen (77 K), the threshold current decreases by more than one order of magnitude. Cooling increases the radiative efficiency of the GaAs diode laser by more than seven times its room-temperature value and permits a larger average output power, since the duty factor (percent of time the laser is on) can be increased by a factor of 40. In addition, cooling increases the bandgap of GaAs, shifting the output from the 900 nm region to 850 nm, where detectors are more sensitive. In any application, one must look at the balance between increased power and efficiency and the additional expense and maintenance of a cooling system before deciding between room-temperature and cryogenic operation of the diode.

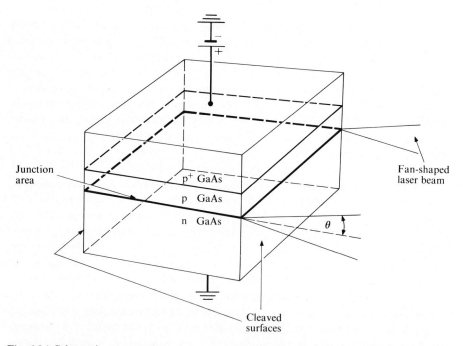

Fig. 6.24 Schematic construction of a semiconductor laser. The emission is confined to the junction region. The narrow width of the junction causes the large half-angle beam divergence θ.

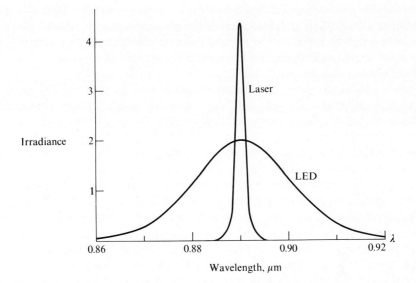

Fig. 6.25 Emission spectrum of a semiconductor laser compared to that of an LED operating below threshold.

All the emission occurs in the narrow junction region. Because of the confinement of the beam to a very small region, diffraction of the light results in a large beam divergence. Light of wavelength λ passing through a slit of width a will be diffracted through an angle given by the half-angle diffraction formula:

$$\sin \theta \approx \frac{\lambda}{a}. \tag{6.7}$$

For a typical semiconductor laser, such as gallium arsenide, $\lambda = 900$ nm and $a = 6\ \mu$m:

$$\sin \theta = \frac{900 \times 10^{-9}}{6 \times 10^{-6}} = 0.15 \quad \text{or} \quad \theta = 8.6°. \tag{6.8}$$

This angular spread is quite large when compared to that of other types of lasers. The fan beam is characteristic of all semiconductor lasers and presents a drawback for some types of applications.

The construction of a simple semiconductor laser was shown in Fig. 6.24. Lasers are made by the growth of an n-type layer on a p$^+$ substrate or by the growth of a p$^+$ layer on an n-type substrate. (p$^+$ material is heavily doped p-type material that serves to provide good ohmic contacts.) Heat treatment is required during or after growth to diffuse some of the p-type dopant into the n-type region to create a lightly doped p-type region. The result is the formation of a p^+-p-n structure. Electrons are injected from the n-region into the p-region where they combine with holes and emit radiation.

Semiconductor lasers that are fabricated from a single semiconductor, e.g., gallium arsenide (GaAs), are called *homojunction lasers*. Homojunction lasers suffer some disadvantages compared to *heterojunction lasers*, which are fabricated from more than one material. In both types of lasers, both the injected electrons and the emitted light must be confined to the junction region for an efficient stimulated-emission process. In a homojunction laser, the confinement of the light to the junction region is a consequence of the free electrons and holes that are present there. They serve to increase the refractive index so that the light is internally reflected rather than transmitted out of the junction region. Although this confinement mechanism works sufficiently well to allow homojunction lasers to operate, the threshold current density is high and the efficiency is low. Some light spreads out of the junction region, reducing the amount available to produce stimulated emission. The electrons are injected to various distances into the p-region before they recombine.

A much higher lasing efficiency and a much lower threshold current density is obtained when two materials are used to form the junction. Materials such as gallium arsenide and its aluminum admixture, gallium aluminum arsenide (GaAlAs), have different refractive indexes and band gaps. The discontinuity in the refractive indexes causes radiation generated within the junction to be reflected back into the region, giving a higher lasing efficiency. The band-gap difference confines the carriers to the junction region and thereby reduces the threshold current density. Heterojunction lasers are of two principle types: *single heterostructure* and *double heterostructure*. The single heterostructure is formed by depositing a p^+-type $Al_xGa_{1-x}As$ layer (x is a positive fraction less than one) on an n-type GaAs substrate. The double heterostructure consists of a triple structure of p-type GaAs sandwiched between p^+-type and n-type $Al_xGa_{1-x}As$. The triple structure is frequently mounted on an n-type GaAs substrate, with p-type GaAs on top to provide an ohmic contact.

The action of the double heterojunction is illustrated in Fig. 6.26, which shows the band structure for a junction composed of n-type $Al_xGa_{1-x}As$ and p-type GaAs and for a junction compound of p-type GaAs and p^+-type $Al_xGa_{1-x}As$. For the $p-p^+$ heterojunction, the increase in the band gap of $Al_xGa_{1-x}As$, as compared to GaAs, produces a potential barrier that reflects injected electrons. The n–p heterojunction prevents holes from crossing into the n-type region, which would prevent electrons from being injected into the junction area. This structure effectively widens the junction and confines recombination radiation to

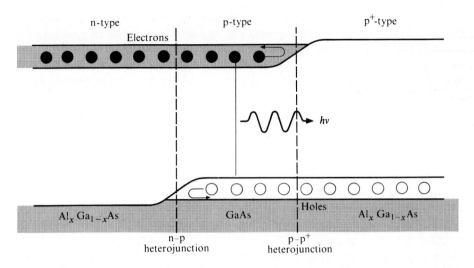

Fig. 6.26 Energy-band diagram for a double-heterojunction laser. The n–p heterojunction at the left creates a barrier against hole transport into the n-region. The p–p$^+$ heterojunction at the right creates a barrier against injected electrons in the p$^+$-region. Both junctions, however, allow easy transport of the other carriers into the junction region.

this region. This double heterostructure approach reduces the threshold current density to about 2000 amperes/cm^2, as compared to 8000 amperes/cm^2 for single heterostructure lasers and to 40,000 amperes/cm^2 for homostructure lasers. Coupled with this decrease in threshold current density is a corresponding increase in power efficiency due to the stricter confinement of the radiation to the active region. Some commercially available heterostructure junction lasers operate continuously at room temperature. One such laser has a 5-mW output at 820 nm with a threshold current of less than 300 mA.

6.4 DYE LASERS

Although they are not as common as either gas lasers or solid-state lasers, liquid lasers possess unique advantages for certain applications. The output of a doped-insulator laser depends greatly on the degree of perfection of the particular crystal used. For example, inhomogeneities in the crystal structure can severely limit the coherence of the laser. The crystals are subject to permanent damage and, once prepared, their dopant concentrations are fixed. Gas lasers do not suffer from these particular difficulties, but have a much smaller concentration of active lasing atoms, simply because the atoms are much further apart on the average. The advantages of liquid lasers are (1) a significantly higher concentration of active atoms that can be varied easily, and (2) a laser medium that is cheap and relatively damage-free and that is not as bulky nor as difficult to handle as a gaseous system.

Of the various types of liquid lasers, the dye laser is the most significant. The dye laser was discovered in 1965 by P. Sorokin and coworkers at the IBM laboratories during an investigation of certain dyes used as saturable absorber Q-switches for ruby lasers. The single most important advantage of the dye laser over other lasers we have studied is that the output is tunable over a significant frequency range. The typical gas or solid-state laser can be tuned only within a very small range (essentially the width of the gain profile). Although the available gas and solid-state lasers cover a large number of discrete wavelengths ranging from the near ultraviolet to the far infrared, there remain substantial portions of the optical band that are not covered by these lasers. A tunable device capable of providing the extremely narrowband, highly coherent light of a laser throughout the optical band is highly desirable in many applications, such as spectroscopy, molecular dissociation, chemical reactions, and isotope separation.

The Active Medium

The active medium of a dye laser consists of an organic dye dissolved in a solvent. When the dye is excited by an external source of short-wavelength light, it emits radiation at longer wavelengths, or *fluoresces*, absorbing a photon at the excitation wavelength and subsequently emitting a photon at the fluorescence wavelength.* The energy difference between the absorbed and emitted photon is accounted for by a nonradiative transition in the dye, ultimately showing up as heat. The absorption and emission spectrum of a typical laser dye is shown in Fig. 6.27. The fluorescence curve in this example, extending through much of the red and yellow portions of the spectrum, is shifted to longer wavelengths compared to that of the absorption curve in the blue and green. This is similar to our discussion of ruby lasers, where the flashlamp is centered in the green and blue portion of the spectrum and the ruby crystal fluoresces in the red. The essential difference here is that the dye fluoresces over an extremely broad band of visible frequencies,

---

* The property of fluorescence, the absorption of light at one color and its reemission at a different color, is found frequently in nature. The blossoms of the azaela and bougainvillaea plants are striking examples of fluorescence. Day-glo signs are another example. There are other everyday materials that exhibit this property on a more subtle level. The whitish glow seen in tonic water is an example. Even this liquid, when illuminated by ultraviolet light, will undergo lasing action. It has been reported that the addition of ethyl alcohol to the tonic water enhances the output. Thus vodka and tonic becomes the world's first drinkable laser.

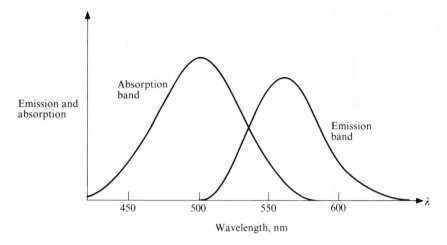

Fig. 6.27 Absorption and emission (fluorescence) spectrum of a typical laser dye.

in contrast to the very narrow fluorescence band of typical solid-state lasers.

The broad fluorescence spectrum of the dye can be explained by the energy diagram of a typical dye molecule, as shown in Fig. 6.28. The dye molecule has two groups of states: the singlet states (labeled S_0, S_1, and S_2) and the triplet states (labeled T_1 and T_2). The singlet states occur when the total spin of the excited electrons in each molecule is equal to zero. The triplet states occur when the total spin is unity. As we noted in our discussion of selection rules and transition lifetimes (Section 3.1), singlet-triplet and triplet-singlet transitions are far less likely than transitions between two singlet or between two triplet states. The dye laser is pumped by the absorption of photons, which excite the molecules from the ground state S_0 to the first excited state S_1. There is a very rapid decay by a nonradiative process to the lowest energy states of S_1. The laser transition occurs between a state near the bottom of S_1 and an intermediate state in S_0. Since there are many vibrational-rotational sublevels within S_0 and S_1, indicated by the multiple lines in the figure, the resulting emission line is very broad. The triplet states T_1 and T_2 are not involved directly in the laser action, but they have a pronounced effect, nevertheless. There is a small probability that the forbidden transition $S_1 \rightarrow T_1$ (called an *intersystem crossing*) will occur. Since transition $T_1 \rightarrow S_0$ (*phosphorescence*) is also forbidden, the molecules tend to pile up in the T_1 state. But the transition $T_1 \rightarrow T_2$ is allowed and, unfortunately, the range of frequencies for that transition coincides almost exactly with the range of laser transition frequencies. Once a significant fraction of the molecules have made

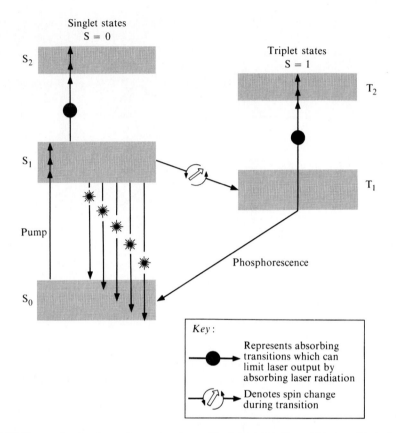

Fig. 6.28 Energy-level scheme for a dye laser. Singlet-triplet ($S_1 \rightarrow T_1$) transitions lead to strong absorptions ($T_1 \rightarrow T_2$) at the laser transition wavelengths, quenching laser action.

the $S_1 \rightarrow T_1$ transitions, the $T_1 \rightarrow T_2$ absorption quickly reduces the laser gain and can quench laser action. For this reason, some dye lasers operate on a pulsed basis with pulse durations shorter than the time that it takes for the population of the T_1 state to reach a significant value. There also may be absorption associated with the transitions to higher singlet states ($S_1 \rightarrow S_2$) for certain dyes, so one must choose dyes for which the frequencies of these transitions do not lie in the region of interest.

The broad tuning range made available by using different dyes as active mediums enables one to cover almost the entire visible range, as shown in Fig. 6.29. It is also evident from the figure that rhodamine 6G is specified in many systems because of its high efficiency (~ 20 percent) and its broad tuning range. The maximum power output of dye lasers depends on the solvent used and on the precise wavelength of the pump laser, as well as on such factors as the type of dye and the alignment of the optical cavity. Certain additives, such as cyclo-

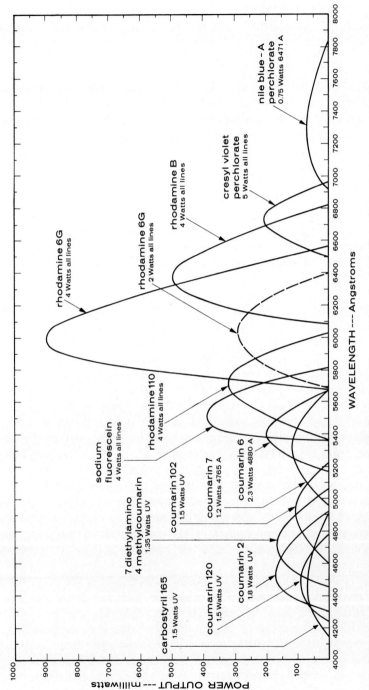

Fig. 6.29 Dye laser output curves of some common laser dyes. Figures below the dye indicate the typical pump power from an argon ion laser required to achieve the tuning curves shown. (Courtesy of Coherent Radiation, Inc.)

octatetraene (COT), can slightly shift the fluorescence of the dye and increase the power output.

Pumping

All dye lasers are optically pumped. The primary requirement is that the pump light source must have an output near the peak of the dye absorption band. By the nature of the dye, the laser outputs occurs at wavelengths longer than that of the excitation source. Rhodamine 6G, for example, which fluoresces around 590 nm (orange region of the spectrum), is pumped by light in the 490–515 nm (blue-green) region of the spectrum. For a dye fluorescing in the deep blue, a pump source with ultraviolet output is required.

The type of pump source dictates not only the laser output range, but also the specific pumping configuration to be used. The first and simplest method of pumping a dye is to treat it as if it were a solid-state crystal rod that is being optically pumped. A tube containing flowing dye is placed in an elliptical reflector and pumped by a linear flashlamp. Flashlamp pumping yields peak powers of several kilowatts and average powers on the order of 50 mW in the visible region.

It is also possible to pump dyes using a nitrogen laser as a source. The dye is optically pumped in a transverse geometry—with the fan-shaped beam from the pump laser focused into a region along one side of the dye cell. The output windows on the ends of the cell may be either flat and antireflection-coated or tilted at the Brewster angle. The mirrors are external to the dye cell, so that a change of tuning range involves only a change of dye cell and a reorientation of frequency-selective devices located in the laser cavity. Because of the short wavelength of the nitrogen laser and its high peak power, dye-laser output can be achieved over a wide spectral region extending from 350–680 nm. In the case of such short pumping wavelengths, one sometimes employs a two-step pumping process. A dye that absorbs the N_2 output at 377 nm is mixed with the lasing dye as an impurity. The impurity dye absorbs the N_2 output efficiently and fluoresces at a longer wavelength for which the host dye is highly absorptive. The latter can now lase due to the secondary excitation. The output pulse energy of a typical nitrogen laser is about 1 millijoule (100 kW peak for 10 nsec); the pumped dye laser output for these excitation levels varies from 2 to 200 microjoules, typically 50 microjoules.

Another widely used pumping source for a dye laser is the powerful blue-green lines or the ultraviolet lines of an argon ion laser. For many dyes that lase at wavelengths greater than 560 nm (yellow through red), the dye absorption bands cover the visible output of the argon laser. A dye, such as rhodamine 6G, can absorb nearly all of the visible output and convert more than 20 percent of the input energy into coherent output at the peak of its emission band. Other dyes, such as coumarin 6, which has a tuning range from about 520 to 560 nm, are pumped by a single argon ion laser line at 488 nm. Dyes that lase in the blue region of the spectrum must be pumped by an argon ion laser whose tube-current characteristics and output mirrors are designed specifically for high ultraviolet output. With multiwatt ultraviolet pumping, the researcher can now obtain a tunable output power level of several hundred milliwatts of blue light—a power level that was available only at discrete wavelengths prior to development of the dye laser.

Because of the extremely high small-signal gain of most laser dyes, only a small amount of active medium is needed. However, the intense absorption and subsequent heating of the small volume of dye plus the rapid buildup of the triplet-state population necessitates a continuous and rapid change of the pumped volume. Failure to do so produces a heated dye that decomposes, resulting in increased absorptive losses for the system. The dye flow can be confined between two windows through which the pump beam and the lasing radiation are transmitted. It has been found through experience, however, that the dye molecules decompose and become attached to the windows (in effect, the laser beam "cooks" the dye onto the windows). To avoid this problem, dye lasers are built around a liquid jet design. The dye volume intersects the pump beam in the form of a smooth laminar sheet of dye expelled into the air through a precisely shaped nozzle. Figure 6.30 shows the nozzle and the cup that catches the dye for return to the dye reservoir. To keep the smooth dye stream from breaking up upon leaving

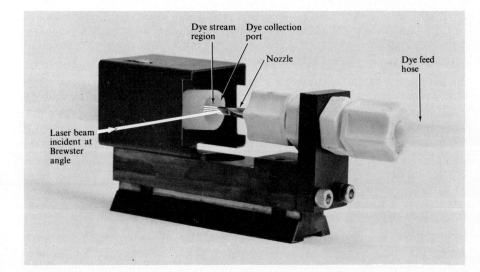

Fig. 6.30 Nozzle assembly of a laminar-flow dye laser. (Courtesy of Spectra Physics, Inc.)

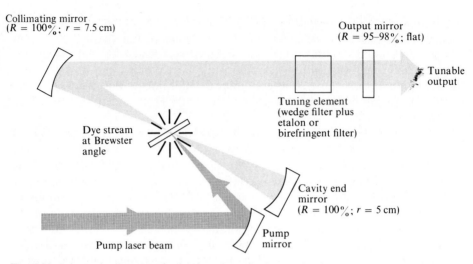

Fig. 6.31 Schematic diagram of a laminar-flow dye laser. The dye-laser cavity is formed by the reflector (radius 5 cm) and the output coupler. The other reflector (radius 7.5 cm) serves to fold the cavity so that the dye-laser output is parallel to the input pump beam. Dye stream flow is perpendicular to page.

the nozzle, ethylene glycol, the basic ingredient in most antifreeze, is frequently used as a solvent because of its high viscosity. The complete configuration is illustrated schematically in Fig. 6.31. The pump laser beam is focused into the dye jet, where it is almost totally absorbed. Any pump light transmitted by the jet is absorbed by the pump stop. Stimulated emission occurs throughout the small pumped volume, but feedback is confined to a direction at a small angle to the pump beam. The cavity is folded into two parts. One part, consisting of the 100-percent reflecting end mirror and the folding mirror, is oriented at an angle to the pump beam; the second part, consisting of the folding mirror and the output coupler (which transmits 2–5 percent of the light), is parallel to the pump beam. Besides returning the output direction parallel to the input beam direction, the folding mirror increases the length of the cavity, thereby decreasing the longitudinal mode separation, introducing more modes beneath the gain curve and increasing the output power over that obtained with a shorter, unfolded arrangement.

Tuning the Output

Tuning the laser output is accomplished by means of a wavelength-dispersing element such as the prism wavelength selector described in Section 5.1. In some cases, a multiple Brewster prism configuration is used for increased dispersion. In other cases, the gain of the dye is so high that a diffraction grating can serve as a combined end mirror and dispersive element. For a grating whose normal is set at an angle θ to the optic axis of the laser, a very narrow band of wavelengths satisfying the relationship

$$m\lambda = 2d \sin \theta, \qquad m \text{ an integer} \tag{6.9}$$

are diffracted back down the laser cavity. The grating order m is usually unity, and d is the distance between rulings of the grating. Other wavelengths are not returned to the cavity and experience high loss. The wavelength can be changed by merely rotating the grating to change the angle θ.

There are a number of other tuning elements that can be inserted into the dye laser cavity. One type used in commercial devices is a *wedge filter*. This filter consists of a substrate with three dielectric coatings. The first and third coatings form highly reflective mirrors; the middle coating is a wedge of dielectric material that provides a variable mirror separation. Together the coatings form a variable-width Fabry-Perot interferometer. By moving the wedge in a direction perpendicular to the cavity axis, the peak transmission of the filter is tuned to a different wavelength. The reduced loss at the new wavelength admits stimulated emission there. The breadth of the laser linewidths obtained by this method may still be too broad for some applications, and one or more plane-parallel etalons similar to those used in single-moding some lasers (see Section 5.2) are inserted into the cavity to narrow the lasing linewidth still further.

Another device that can be used to tune a dye laser is a novel *birefringent filter* consisting of several quartz waveplates (Section 2.7) of different thicknesses. These plates are placed in the laser cavity at the Brewster angle such that the vertically polarized light in the cavity experiences no loss by reflection at the plate surfaces. As we pointed out earlier, the no-loss condition permits amplification of light in the cavity with a high degree of linear polarization. To understand the operation of the tuning plates, consider the effect of one such plate on light in the cavity. The crystal axes of the quartz are oriented such that the plate behaves as a full waveplate for vertically polarized light if λ_0, the wavelength in a vacuum, satisfies the relation $d(n_{\text{slow}} - n_{\text{fast}}) = m\lambda_0$ (m an integer), where d is the distance traveled by the light in the plate. For other wavelengths, transmission of the vertically polarized light through the plate results in elliptical polarization. After reflection at an end mirror, this elliptically polarized light experiences loss by reflection at the next encounter with the waveplate surface. This loss prevents lasing at wavelengths that differ much from those satisfying the full-wave condition. Although a single thin plate can isolate one band within the lasing region, the resulting linewidth is approximately 0.3 nm, a value somewhat large for most applications. However, if a second waveplate twice the thickness of the first is also placed in the cavity, the linewidth is reduced to 0.1 nm. For some applications, a third plate twice the thickness of the second is inserted to further reduce the linewidth to less than 0.03 nm.

Tuning the laser is accomplished by rotating the plates, which are mounted on a common stage, about the normal to the plate surfaces, as shown in Fig. 6.32.

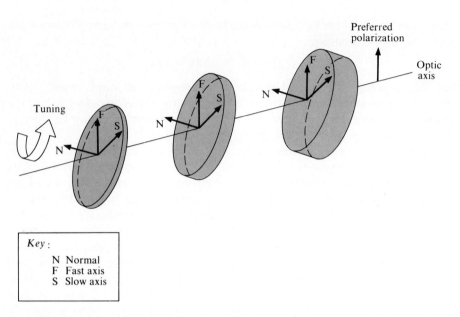

Fig. 6.32 Schematic of a birefringent filter. The thicknesses of the plates are in the ratio 1:2:4. Tuning is accomplished by rotating the plates together about the normal to the plates.

Because the plates are inclined to the optic axis, the rotation effectively changes the slow axis refractive index from n_{slow} to n'_{slow}, thereby changing the preferred wavelength to $\lambda'_0 = d(n'_{slow} - n_{fast})/m$.

The safety precautions to be taken with the dye laser are essentially the same as with any high-power, visible-light laser. When the dust cover is off the dye laser, a number of beams are present at odd angles. During alignment or inspection, one should be aware of the directions of all beams and take care to block them or avoid them. One problem with dye-laser safety is that one's goggles must block all the argon pump lines as well as the entire tunable range of some dyes. In some cases, this criteria could only be met by totally opaque goggles!

Isotope Separation

One of the most significant and interesting future applications of the dye laser may be that of isotope separation, particularly in the case of uranium. Natural uranium is composed of two principal isotopes, U^{235} and U^{238}, with U^{235} accounting for less than one percent. For use as a fuel in nuclear electric power plants, a mixture with at least 3 percent U^{235} is required. The current method of uranium-isotope separation is gaseous diffusion, which is a very expensive and time-consuming process. The proposed separation mechanism of uranium using dye lasers is based on the fact that the dye laser can be so precisely tuned that the beam can selectively excite the U^{235} atoms without exciting the U^{238} atoms. The excited U^{235} atoms are then ionized, probably by another light source of short wavelength. The charged U^{235} atoms can then be separated from the uncharged U^{238} atoms by an electrostatic field. Although the technology for large-scale uranium-isotope separation has not yet been developed, feasibility experiments have proved successful. For a large-scale isotope-separation system, a laser design has been proposed that will provide an average power of 10 W at 500 nm. The pulse repetition rate will be on the order of 400 Hz with a duration of about one microsecond. Instead of using visible-light lasers, some isotope separation schemes require tunable infrared lasers to interact with the molecular vibrations of uranium hexafluoride (UF_6). Research is being done to develop broadband tunable coherent sources in the region from 2 to 20 μm.

6.5 EXTENDING THE RANGE

One of the major goals of laser scientists and engineers has been to develop sources of coherent light that can be tuned over the entire optical spectrum, from the far infrared through to the ultraviolet and beyond. The dye laser is an especially important development in this regard, for its output is tunable over a range that exceeds the limits of the visible spectrum. There are still substantial gaps in the laser spectrum, however, regions where known laser transitions are sparse and tunable only over narrow bandwidths. The broad fluorescence transitions that serve as the basis for the tunable dye laser are not found in the far infrared portion of the spectrum, and otherwise usable laser dyes are quickly destroyed by the intense pump radiation when pumped for laser output in the ultraviolet.

Nonlinear Optics

In their quest to fill these gaps, many laser scientists have exploited *nonlinear effects* in certain optical materials. In 1961, researchers at the University of Michigan focused the 694.3-nm light from a ruby laser onto a crystal of quartz and detected at the output not only the ruby laser light itself but also light at twice the frequency, i.e., at a wavelength of 347.2 nm. Although much weaker than the 694.3 nm light, this short-wavelength radiation was nevertheless characterized by the monochromaticity and spatial coherence of laser light. The process of generating this light, known as *frequency doubling* or *second harmonic generation*, is only one example of a host of nonlinear optical effects that have been used to extend the tunable range of lasers.

Second harmonic generation is often used to double the infrared output ($\lambda = 1.06$ μm and other lines) of neodymium lasers to the yellow-green region of the spectrum (e.g., $\lambda = 530$ nm), where there are few strong laser lines available. Harmonic generation also can be used for tripling the frequency of laser light. The nonlinear characteristics of rubidium and other alkali metals, for example, have been exploited to frequency-triple the output of a neodymium laser up to 353 nm, in the ultraviolet. In theory, higher-order conversion processes are possible, but the diminishingly small coefficients associated with higher-order effects make these processes impractical for most applications.

The possibilities for generating coherent light at new frequencies are not limited to the generation of harmonics. It is also possible to "mix," or combine, two waves to produce new waves at frequencies equaling the sum and difference of the constituent frequencies and their harmonics. One such process is *parametric amplification*. In this process, a lower-frequency wave at v_2, referred to as the *signal wave*, mixes in the nonlinear crystal with a higher-frequency wave at v_1, which is called the *pump wave*. As the two waves propagate through the medium, the signal wave is amplified at the expense of the pump wave, growing in amplitude as the pump-wave amplitude decreases. At the same time, a third wave at $v_3 = v_1 - v_2$, known as the *idler wave*, is generated. If the nonlinear crystal is placed inside an optical resonator that is resonant at the signal frequency or at the signal and the idler frequencies, then laser-like oscillations can occur at these frequencies. The device, which bears a strong resemblance to a laser, is known as a *parametric oscillator*.

The parametric amplification process would be useful even if it were limited to the generation of coherent radiation at frequencies equaling the difference of two existing laser frequencies. In fact, parametric oscillators are capable of producing coherent radiation that is tunable over most of the optical spectrum. The reason is that the signal wave, which is amplified in the oscillator, does not need to be input to the device but can originate with the noise photons (thermal noise) that are always present in the crystal. These noise photons have a wide range of frequencies, predominantly in the infrared. For a particular crystal temperature and orientation to the pump beam and resonator axis, only one signal-wave frequency will be amplified. To tune the device, it is only necessary to vary the temperature of the crystal or change its orientation. The useful output may be either the signal wave or the idler wave, depending on the range of output frequencies desired. Rapid tuning over a limited frequency range can also be achieved by electrooptic variation of the refractive indexes of the crystal. As with a laser, there is a threshold pump level that must be exceeded for steady-state oscillations. Most parametric oscillators use a visible laser, such as an argon laser, or the second harmonic of a neodymium laser as the pump source and produce tunable infrared radiation as the signal. The idler is also in the visible and has its frequency shifted too.

6.6 LASERS TO COME

Besides shifting the laser output frequency to other regions of the spectrum by nonlinear techniques, researchers are also actively pursuing the development of entirely new types of lasers. The coherent generation of soft x-rays ($\lambda = 10^{-8} - 10^{-9}$ m) is a particularly important area of research.

There are two principal obstacles to the development of x-ray lasers: Reflectors for x-ray wavelength optical cavities are virtually nonexistent, and pumping becomes increasingly difficult with shorter wavelengths. Because any solid material absorbs x-ray radiation, it is not possible to construct a conventional laser cavity to create feedback, and little hope is offered for the possibility of deviating x-rays into a circular path by a series of glancing reflections from crystals. Thus, coherent x-ray devices probably will be laser amplifiers rather than laser oscillators. The other serious problem, that of pumping these lasers, arises from the extremely short lifetimes of x-ray transitions, which are on the order of 10^{-15} sec. Maintaining population inversion requires that pumping powers on the order of one watt per atom be used! Pumping systems using picosecond pulses from a master oscillator and an amplifier chain (discussed in Section 9.3) that can produce terawatt (10^{12} watts) peak powers are becoming available, in addition to the electron-beam pumped excimer systems discussed in Section 6.1. If an x-ray laser could be constructed, it could be used for short-exposure radiography, examination of dense gases at high temperatures, and submicroscopic holography.

There also have been suggestions and preliminary investigations of techniques for developing *grasers* (gamma ray lasers) having output wavelengths in the vicinity of 0.1 nm. Operating without mirrors, the grasers would amplify radiation by stimulated emission from excited nuclear states.

Where will it end? The only fundamental limit that has been suggested is that of *pair production*, an event where a photon has sufficient energy to decay spontaneously into an electron and a positron. Such a laser operating at the threshold energy for pair production would have a wavelength of 1.2 picometers (1 picometer $= 10^{-12}$ meter). By comparing the frequency of such a laser to that of the ammonia maser at 24 GHz, we see that devices employing amplification by stimulated emission would operate over ten decades of frequency.

PROBLEMS

6.1   Referring to Fig. 6.17, show that the angle of cut of the Brewster angle laser rod is θ_B, and that the angle of tilt of the rod with the vertical is $180° - 2\theta_B$. Begin with the fact that the angle of the light with the normal to the rod face (external to the rod) is θ_B, the Brewster angle. (The other condition is that the light must travel parallel to the surface of the rod in order to use all the available active medium efficiently.)

6.2   Estimate the divergence angle of a semiconductor laser, if the junction region were made greater than ten times larger than that of the text example, to about 60 μm.

6.3   The index of refraction of gallium arsenide is 3.4. If the free-space wavelength is 900 nm, calculate the wavelength in the crystal. Calculate the reflectivity at the semiconductor-air interface for light incident from within the crystal. The reflection coefficient at normal incidence at an interface is

$$R = \left(\frac{n_1 - n_2}{n_1 + n_2}\right)^2$$

where n_1 and n_2 are the refractive indexes of the two mediums. Assume $n_{air} = 1$.

6.4 A pulsed GaAs semiconductor diode laser requires a rectangular current pulse of 50 amperes to produce a 5-watt peak output pulse of 200 nsec duration at 25°C. The pulse repetition rate is 5000 pulses per second. During the current pulse, the voltage drop across the diode is 0.2 volts.

  a) Calculate the power efficiency of the laser.
  b) What is the duty cycle (time on/total time) of the laser?
  c) What are some of the reasons for less than 100 percent efficient operation?

6.5 The *external quantum efficiency* of a diode laser is given by

$$\eta_{qe} = \frac{N_v}{N_e}$$

where N_v is the number of photons per second of frequency v emitted and N_e is the number of electrons per second that flow into the diode. The power efficiency of the diode is given as $\eta_P = W/IV_d$, where V_d is the voltage drop across the diode, W is the emitted laser power, and I is the diode current.

  a) Show that $\eta_{qe} = \dfrac{e\lambda W}{hcI}$.

  b) Using the result obtained in (a) and Eq. (6.6), show that $\eta_P = \eta_{qe}V_g/V_d$.
  c) What is the power efficiency for a diode laser with an output at 900 nm, if $V_d = 2$ volts, and $\eta_{qe} = 15$ percent?

The following six problems have a common set of instructions. You are to select the necessary laser components from a common equipment list and specify the transmission curves of some of the components. Finally, add the lengths of the components placed between the mirrors to obtain a laser length, then calculate the g values for each mirror and evaluate the stability of the laser.

    List your selection of components, left to right, on a series of lines, starting from the 100-percent reflecting mirror on the left with the output beam traveling to the right. Those components with a parenthesis as part of the symbol require the appropriate transmission curve to be given. (Components may be used more than once.) Off-axis components should be listed after the on-axis component nearest to it (e.g., CR3 (FL)).

Example: A high-power, continuous-wave laser for use in material applications ($\lambda = 10.6\ \mu m$).

Answer: M(2)   CO2   M(1)   →   Output

Lengths: 90 cm

$$L = 90; \qquad g_1 = 1 - \frac{90}{110} = g_2 = 0.18; \qquad g_1 g_2 = 0.033$$

Note that M(3) will not work because of the high transmission at 10.6 μm. M(4) will not work because the g-value would be negative and would have to multiply the g for M(2), which is positive.

6.6   A modulated coherent output ($\lambda = 633$ nm) for information processing.

___ ___ ___ ___ ___ ___ ___ ___   →   Output

6.7   A high-power pulsed visible output ($\lambda = 530$ nm) for atmospheric pollution studies.

___ ___ ___ ___ ___ ___ ___ ___   →   Output

6.8   A tunable visible output ($\lambda = 560-640$ nm) for spectroscopy of biological samples. Use an ultraviolet pump to eliminate background that may be encountered with visible pumping. Include additional intercavity space to ensure a small axial mode separation of 750 MHz.

___ ___ ___ ___ ___ ___ ___ ___   →   Output

6.9   A source for large-area holography that requires large pathlength difference. A blue output ($\lambda \sim 400-500$ nm) is needed for high-speed, blue-sensitive plates.

___ ___ ___ ___ ___ ___ ___ ___   →   Output

6.10   A repetitive, high peak-power pulse output in the visible ($\lambda = 694$ nm) for repairing printed circuits.

___ ___ ___ ___ ___ ___ ___ ___   →   Output

6.11   A source for scanning photographic emulsions (maximum sensitivity at 325 nm) at high speeds. The laser is internally modulated.

___ ___ ___ ___ ___ ___ ___ ___   →   Output

EQUIPMENT LIST

Symbol	Length (cm)	Description
AR+	70	Argon laser tube + power supply
CO2	90	Carbon dioxide laser + power supply
CR3	50	Ruby rod
FD( )	5	Flowing dye stream + pump (Specify trans. characteristics of dye in parentheses.)
HC	30	Helium-cadmium laser tube + power supply
HN	80	Helium-neon tube + power supply
N2	70	Nitrogen laser tube + power supply
ND	50	Nd^{3+}:YAG rod
BF( )	15	Birefringent filter
DC( )	10	Dye cell; low irradiance transmission curve
E01( )	15	Electrooptic crystal with electrodes and high voltage supply
SHG	20	Frequency doubling crystal
ET( )	15	Etalon + oven
(FL)	Off line	Flashlamp
G	0	Grating
(M)	Off line	Magnet
M(1)	0	Mirror; transmission curve #1; radius of curve.:110 cm
M(2)	0	" " " #2; " " " :110 cm
M(3)	0	" " " #3; " " " : 66 cm
M(4)	0	" " " #4; " " " : 66 cm
P( )	5	Polarizer
PR( )	5	Prism
SP( )	Specify	Additional space between mirrors (Problem 6.8). (Put amount in cm in parentheses: 10 cm additional = SP(10).)

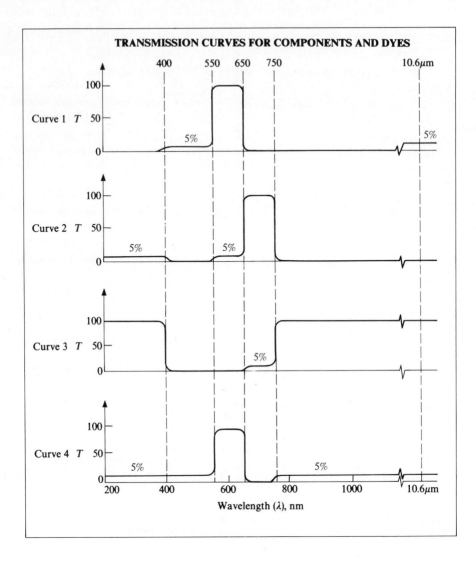

REFERENCES

6.1 Arnold L. Bloom (1968), *Gas Lasers*. New York: Wiley.

6.2 S. S. Charschan (ed.) (1972), *Lasers in Industry*. New York: Van Nostrand Reinhold. The first chapter includes a short review of the lasers discussed in this chapter. Contains additional energy diagrams and a discussion of the water-vapor laser not given in this text.

6.3 Bela A. Lengyel (1971), *Lasers* (2nd edition). New York: Wiley-Interscience.

6.4 *Scientific American* articles:
    Arthur L. Schawlow, "Optical Masers," June 1961.
    Arthur L. Schawlow, "Advances in Optical Masers," July 1963.
    George C. Pimentel, "Chemical Lasers," April 1966.
    Alexander Lempicki and Harold Samelson, "Liquid Lasers," June 1967.
    Peter Sorokin, "Organic Lasers," February 1969.
    William T. Silfvast, "Metal-Vapor Lasers," February 1973.
    Richard N. Zare, "Laser Separation of Isotopes," February 1977.

Holography

Although holography was invented more than a decade before the laser,* the requirement of the holographic process for light with a high degree of temporal and spatial coherence has closely linked the development of holography to the development of the laser. Holography has become a very important area of laser application, encompassing not only eye-catching three-dimensional displays, but also optical memories, optical components, and nondestructive testing of materials. Often referred to as *wavefront reconstruction*, holography provides the scientist, the technologist, and the advertising executive with a means of recreating wavefronts of light just as though they were coming from an object that has long-since been removed.

The basic concept of wavefront reconstruction is illustrated in Fig. 7.1. In part (a), a *holographic plate*, which is a high-resolution photographic emulsion on a glass plate, is exposed simultaneously to waves of light coming from an "object" source and to waves of light coming from a "reference" source—in this case a focused point of light. Both the object waves and the reference waves come from the same laser source. Because of their high degree of mutual coherence, the two waves produce an interference pattern on the plate. This pattern is recorded by the photographic emulsion. The holographic plate is developed, dried, and then reilluminated with the same reference wave, as shown in Fig. 7.1(b). Much of the light passes directly through the hologram. Some of the reference light, however, is diffracted by the interference pattern now preserved in the photographic emulsion. The diffracted wave appears to an observer as though it were coming from the object itself. The wave is a replica, or reconstruction, of the original object

---

* The first hologram was recorded in 1948 by Dennis Gabor, who received the 1972 Nobel Prize in physics for his work.

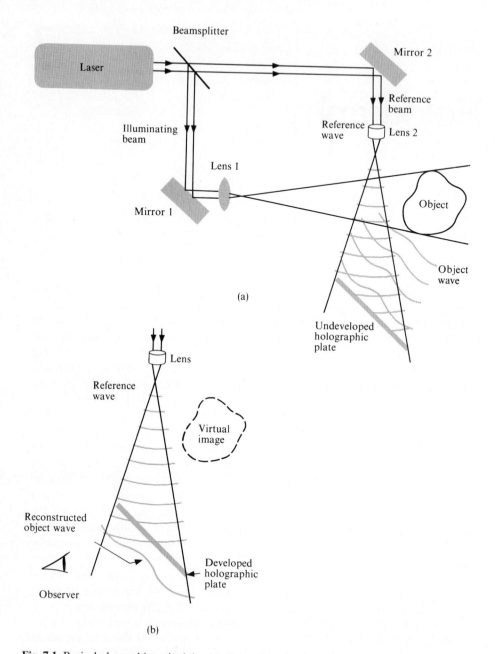

(a)

(b)

Fig. 7.1 Basic holographic principle. (a) Recording the interference pattern produced by interfering object and reference waves. (b) Reconstructing the object wave by illuminating the hologram with the reference wave.

wave, having all the characteristics of that wave. The hologram serves as a window through which the object can be viewed from different elevations and angles, just as though the object were still present. Viewing a well-made hologram, the observer experiences a strong urge to reach behind the plate and touch an object that is not there!

Examples of a reconstructed holographic image are shown in Fig. 7.2. In photographing the reconstructed hologram, the camera was repositioned and refocused between successive photographs.

(a)   (b)

Fig. 7.2 Photographs of a reconstructed holographic image of a model automobile. The camera position and plane of focus were changed between (a) and (b).

Holographic principles are often discussed in terms of complex algebra and phasor representations (see, for example, Chapter 8 of Ref. 7.4). However, the mathematical sophistication of this approach is not necessary for an understanding of the basic principles. Our approach here is based instead on the analysis of a simple optical device, the sinusoidal grating or sine grating, which can be made using a laser and photographic film.

7.1 THE SINE GRATING

Holography requires the recording of an interference pattern on a photographic film or plate. In Section 2.5, we described the sinusoidal interference pattern that is created when two mutually coherent plane waves interfere. A recording of this pattern can be considered to be an elementary form of hologram. In Fig. 7.3, we

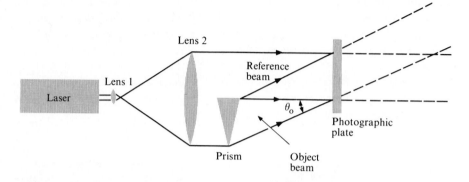

Fig. 7.3 Recording of a sinusoidal grating. The two lenses expand and collimate the beam from the laser. A portion of the collimated beam is deflected upward by the prism to interfere at the photographic plate with the undeflected portion of the beam.

show a simple arrangement of optical components for recording such a pattern. Part of a collimated beam of laser light is deviated by a prism such that two plane waves strike the plate, one with normal incidence, the other making an angle θ_o to the first. Consistent with our earlier discussion, we shall speak of the normally incident wave (the choice is arbitrary) as the reference wave, the other as the object wave. In Section 2.5, we showed that the interference pattern has the form $I(x,y) = CA^2[1 + \cos(kx\sin\theta_o + \Delta\varphi)]$. This result was obtained assuming that the two interfering waves had equal amplitudes. In the more general case of unequal amplitude waves, the irradiance pattern exposing the photographic plate can be shown to have the form

$$I(x,y) = I_o + I_m \cos(kx\sin\theta_o + \Delta\varphi) \qquad (7.1)$$

where I_o is the bias irradiance and I_m is the fringe modulation. These latter two parameters must satisfy the relationship $I_m \leqslant I_o$, since it is always required that

$$I(x,y) \geqslant 0 \qquad (7.2)$$

A typical photographic emulsion has a light transmittance versus-exposure characteristic, or $T - \varepsilon$ curve, similar to the one shown in Fig. 7.4. For short, low-irradiance exposures (left side of the curve), the developed film transmits almost all of the incident light; for long, high-irradiance exposures (right side of the curve), the transmittance is nearly zero and the light is blocked. When exposing the photographic emulsion to the sinusoidal interference pattern, one tries to operate on the linear portion of the transmittance-versus-exposure curve. In this

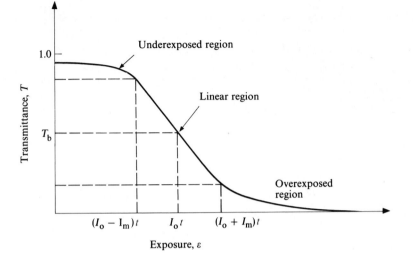

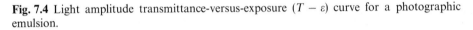

Fig. 7.4 Light amplitude transmittance-versus-exposure $(T - \varepsilon)$ curve for a photographic emulsion.

region, an increase in exposure causes a proportional decrease in T, the trans-mittance of the photographic plate.* The *exposure* on the plate equals the product of the incident irradiance distribution and the duration of the exposure, i.e.,

$$\varepsilon(x,y) = I(x,y)t \qquad (7.3)$$

where $\varepsilon(x,y)$ is the exposure distribution (J/cm^2), $I(x,y)$ is the exposing irradiance distribution (W/cm^2), and t is the exposure duration in seconds. In order to ensure a linear recording of the sinusoidal fringes described in Eq. (7.1), exposure time t must be chosen such that the *bias exposure* (ε_b), given by $\varepsilon_b = I_o t$, is near the

---

* T is the *light amplitude transmittance*, or simply the amplitude transmittance of the film, as distinguished from the light irradiance transmittance used extensively in photographic work. For a photographic transparency, the amplitude transmittance is defined, for every point on the transparency, by

$$T = \frac{E_{\text{out}}}{E_{\text{in}}}$$

where E_{in} is the amplitude of the electric field of the wave incident on the transparency, and E_{out} is amplitude of the electric field of the transmitted wave. See Chapter 7 of Ref. 7.4 for further discussion.

center of the linear portion of the curve, as noted in Fig. 7.4. Further, the fringe amplitude (I_m) must be sufficiently small that the extreme values of the exposure, $(I_o + I_m)t$ and $(I_o - I_m)t$, still lie within the linear portion of the curve. If these conditions are satisfied, the light amplitude transmittance of the exposed and developed plate will have the form

$$T(x) = T_b + M \cos (kx \sin \theta_o + \Delta\varphi)$$
$$= T_b + M \cos \left(\frac{2\pi}{\lambda} x \sin \theta_o + \Delta\varphi\right) \tag{7.4}$$

where T_b is the *bias transmittance*, and M is the *grating modulation*, or *modulation constant*. This recorded pattern of parallel dark and light lines is called a *sine grating*. A magnified portion of such a grating is shown in Fig. 7.5. Equation (7.4) is frequently written in the more convenient form,

$$T(x) = T_b + M \cos (2\pi f_g x + \Delta\varphi) \tag{7.5}$$

where f_g, the *spatial frequency of the grating*, is given by

$$f_g = \frac{\sin \theta_o}{\lambda} \tag{7.6}$$

The spatial frequency is the number of periods of the grating per unit distance in the x- (grating) direction, and is the reciprocal of the grating period, i.e.,

$$f_g = \frac{1}{d} \tag{7.7}$$

where d is the period of the sinusoid, $\lambda/\sin \theta_o$. Figure 7.6 is a plot of the grating transmittance versus x.

Fig. 7.5 Magnified portion of a sine grating.

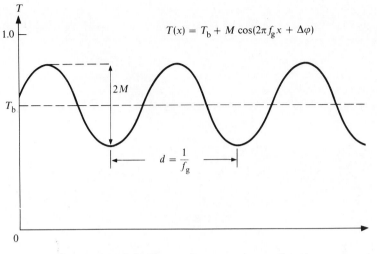

Fig. 7.6 Plot of the sine grating transmittance function.

The sine grating is, in a sense, the simplest of holograms, a recording of the interference pattern associated with a plane reference wave and a plane object wave. The reconstruction, or "playback," of such a hologram is shown in Fig. 7.7, where a collimated beam of light of wavelength λ duplicates the original reference wave. As shown, the incident beam is split into three component beams by diffraction at the sine grating. These beams emerge from the grating with directions given by the grating equation for sinusoidal gratings:*

$$d(\sin \theta_d - \sin \theta_i) = n\lambda \tag{7.8}$$

or

$$\sin \theta_d - \sin \theta_i = n\lambda f_g, \qquad n = -1, 0, 1 \tag{7.9}$$

where θ_d is the angle the diffracted beam makes to the grating normal, and θ_i is the angle the incident beam makes to the normal. For the case of normally incident illumination, $\theta_i = 0$, with the result

$$\sin \theta_d = n\lambda f_g, \qquad n = -1, 0, 1 \tag{7.10}$$

---

* The classical grating formula is the same as Eq. (7.8),

$$n\lambda = d(\sin \theta_d - \sin \theta_i), \qquad d = 1/f_g$$

except that n can assume other integer values. This expression covers the more general case of a grating transmittance function that is periodic but not sinusoidal. A collimated light beam incident on such a grating is split into *multiple* diffracted beams. The beams corresponding to $n = \pm 2, \pm 3, \ldots$ are referred to as higher diffraction orders. In the special case of a sinusoidal grating, only the *zero order* ($n = 0$) and the two *first order* ($n = \pm 1$) *diffraction components* are produced.

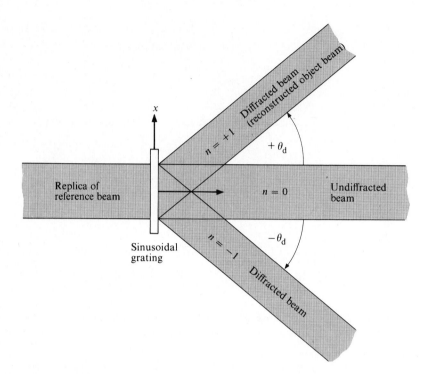

Fig. 7.7 Reconstruction of the object beam at a sine grating. Illumination of the grating with the original reference beam results in two diffracted beams, one of which duplicates the original object beam.

Substituting for n, we obtain the following expressions for the diffracted beams:

$$\sin \theta_d = 0, \qquad n = 0 \qquad\qquad (7.11a)$$

$$\sin \theta_d = \lambda f_g, \qquad n = +1 \qquad\qquad (7.11b)$$

$$\sin \theta_d = -\lambda f_g, \qquad n = -1 \qquad\qquad (7.11c)$$

The first beam, the *undiffracted beam* corresponding to $n = 0$, continues directly on, although with reduced amplitude. The angles at which the two diffracted beams leave the grating can be determined from Eqs. (7.11b) and (7.11c). Of special interest is the $n = +1$ beam, since a comparison of Eq. (7.11b) with Eq. (7.6) shows that the $n = +1$ diffracted beam travels in the same direction as the original object beam, i.e.,

$$\sin \theta_d = \sin \theta_o \qquad\qquad (7.12)$$

or

$$\theta_d = \theta_o \qquad\qquad (7.13)$$

The object wave has, in effect, been reconstructed by reillumination of the grating by the reference wave.

We can also illuminate the grating with a replica of the object wave. In this case, $\theta_i = \theta_o$, and the corresponding expressions for the diffracted beams are, from Eq. (7.9),

$$\sin \theta_d = \sin \theta_o, \qquad\qquad n = 0 \qquad\qquad (7.14a)$$
$$\sin \theta_d = \lambda f_g + \sin \theta_o, \qquad n = +1 \qquad\qquad (7.14b)$$
$$\sin \theta_d = -\lambda f_g + \sin \theta_o, \qquad n = -1 \qquad\qquad (7.14c)$$

Of special interest to us this time is the $n = -1$ term. If we use Eq. (7.6) to substitute for f_g in Eq. (7.14c), we obtain the result

$$\sin \theta_d = -\sin \theta_o + \sin \theta_o = 0 \qquad\qquad (7.15)$$

or

$$\theta_d = 0 \qquad\qquad (7.16)$$

But that is the direction of the original reference wave. Before, illumination of the grating with a replica of the reference wave resulted in a reconstructed version of the object wave. Now we see that illumination of the grating with a replica of the object wave results in a reconstruction of the reference wave. This result should not be too surprising, since our choice as to which plane wave was "reference" and which was "object" was completely arbitrary. The principle is broadly applicable, however, and can be extended to the holographic recording of very complicated interference patterns, as we shall see in Section 7.3.

Reconstructing Points of Light and Continuous Distributions

Not only can beams of light be reconstructed, but focused points of light as well. Consider the recording and reconstruction geometries illustrated in Figs. 7.8 and 7.9. Two mutually coherent point sources, a reference source on-axis and an object source a distance x off-axis, lie in the focal plane in front of a collimating lens. These sources produce two plane waves which interfere at the photographic plate. The angle θ between the waves is given by $\theta = \tan^{-1}(x/f)$, where f is the

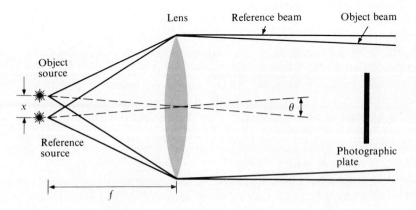

Fig. 7.8 Recording a sine-grating hologram with an object point source and a mutually coherent reference point source.

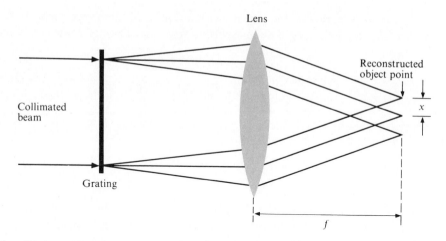

Fig. 7.9 Reconstruction of the original object point of light and its mirror image by illumination of the grating with a replica of the original reference wave.

focal length of the lens. As before, the photographic plate is exposed and developed to produce a sinusoidal grating with grating frequency $f_g = \sin \theta / \lambda$.

When the hologram is reilluminated with a normally incident plane wave (a replica of the reference wave), as shown in Fig. 7.9, the transmitted waves are focused to points of light in the focal plane behind the lens. The point on-axis, corresponding to the reference point source, would appear even without the grating. The other two points, however, result from diffraction of the reference wave at the hologram and correspond to a replica of the original point object source and its mirror image. The separation x, given by $x = f \tan \theta$, is consistent with the recording geometry.

Simple sine-grating holograms of *multiple* points of light can also be recorded and reconstructed. Figure 7.10(a) shows, for example, an arrangement of three noncolinear object points, which replaces the single object point source of Fig. 7.8. When light from these object sources is collimated, plane waves traveling in three different directions interfere with the reference wave to produce three sinusoidal irradiance patterns at the holographic plate, each with a different spatial frequency and orientation. The developed plate is a composite grating similar to the one shown in Fig. 7.10(b). When this collection of superimposed gratings is reilluminated with the reference wave, as was done with the single grating in Fig. 7.9, the focused pattern of light in the back focal plane of the lens has the appearance shown in Fig. 7.10(c). The object distribution has been re-created in its original position relative to the reference point. On the other side of the reference point is created a mirror image of the object, generally referred to as the *conjugate image*. If the original object sources have differing amplitudes, the individual recorded grating patterns have correspondingly different modula-

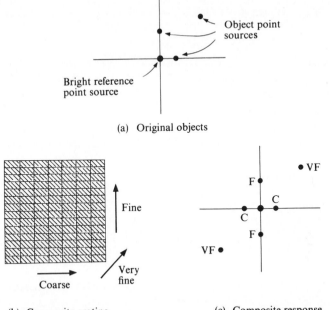

(a) Original objects

(b) Composite grating   (c) Composite response

Fig. 7.10 Recording and reconstruction of a multiple point-source object sine-grating holo-gram. Light from coplanar, mutually coherent point sources shown in (a) is collimated by a lens. The resulting interference pattern recorded on the holographic plate is the composite grating shown in (b), with three sinusoidal components. The reconstructed light distribution has the appearance shown in (c).

tion constants (see Eq. 7.4). These, in turn, establish the relative brightnesses of the reconstructed points of light to match those of the original object source points.* A continuous object—for example, an alphabet character or logo, as shown in Fig. 7.11—can be treated as a collection of a large number of discrete

---

* It should be pointed out that in the recording of such a composite grating, there will be interference at the holographic emulsion between waves from the different object points as well as between the object waves and the reference wave. The resultant "object-object" in-terference patterns produce additional grating components in the processed emulsion that are not desirable. These can be ignored if the amplitude of the reference wave is large com-pared to the amplitudes of the object waves. (Practically, a factor of three is sufficient.) Under these circumstances, the object-reference interference patterns have a much higher contrast than interference patterns arising from different combinations of object point waves, and the desired waves reconstruct with comparatively greater amplitude than the undesired waves. The object-object interference patterns can also be ignored if the object sources are confined to a small region and are well separated from the reference source. Object-object interference patterns then have low spatial frequencies and, on reconstruction, produce a distribution of light that is confined to a small region about the reference point.

Fig. 7.11 A large and very complicated array of gratings could be generated to produce a diffraction response of the kind shown here.

point sources, each one merging with its neighbor. Each of these individual sources creates its own grating on the film. When reconstructed, the focused spots of light corresponding to the various grating components merge and produce a recognizable image.

The composite grating-type hologram is useful when planar objects and images are of interest, such as a page of printed material or a photograph. The full three-dimensional images commonly associated with holography cannot be produced with this kind of hologram. Nonetheless, the grating-type hologram exhibits features characteristic of the great majority of holograms. To begin with, the image information has been recorded not in image form but in a *coded* format. In this case, each point in the image distribution is encoded as a separate sinusoidal component of the composite grating transmission distribution. Second, image information is recorded with *spatial redundancy*: Each sinusoid extends over the entire hologram. The incident light beam need therefore illuminate only a portion of the hologram and the image will still appear.

A third very important feature of the sine-grating hologram, and indeed of holograms in general, is the ability of the holographic recording to preserve information regarding the relative *optical phase* of the light at each object point, not merely the amplitude of the light. Consider again the recording of a sine-grating hologram of the three object points shown in Fig. 7.10. The wave amplitude for each point of light, including the reference, has a time dependency of the form $\cos(\omega t + \varphi_i)$; $i = 0, 1, 2, 3$. In particular, the object points have temporal phases $(\varphi_1 - \varphi_0)$, $(\varphi_2 - \varphi_0)$, and $(\varphi_3 - \varphi_0)$ relative to the phase of the reference point source, φ_0. When the hologram is recorded, these relative temporal phases determine the relative spatial phases of the corresponding sinusoidal components of the hologram; i.e., they determine the positions of the individual grating components. When the hologram is reconstructed, these grating phases determine, in turn, the relative temporal phases of the three reconstructed object points of light. The holographic process is thus not limited simply to the recording and reconstruction of an object radiance distribution; it allows, rather, the reconstruction of a *true replica* of the object light distribution, both amplitude and

---

THE SINUSOIDAL ZONE PLATE

The sinusoidal zone plate, often referred to as a Gabor zone plate, bears a resemblance to the Fresnel zone plate discussed in many optics texts. The Fresnel zone plate is either transparent or opaque, however, whereas the Gabor zone plate exhibits continuous variations in light transmittance values (see the figure below). The sinusoidal zone plate has a light amplitude transmittance of the form

$$T(x,y) = T_b + M \cos 2\pi\alpha(x^2 + y^2),$$

an expression comparable to Eq. (7.5) for the sine grating. The zone plate transmittance varies sinusoidally in the radial direction, but with a "local frequency" that increases linearly with increasing radial distance from the center of the plate. This increasing radial frequency makes the zone plate behave like a lens, or, more properly, like the combination of a positive (converging) lens, a negative (diverging) lens, and a simple attenuator (with no focusing power). An off-axis section illuminated by a collimated light beam is shown. The focal distance of the two lens components is given by $\pm 1/(2\lambda\alpha)$, respectively. A sinusoidal zone plate can be produced by recording the interference pattern from two spherical waves of different radii of curvature.

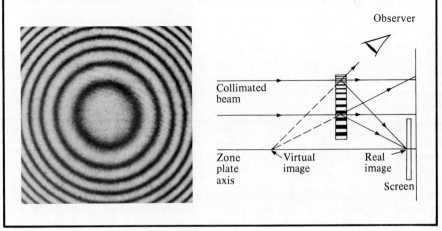

Observer

Collimated beam

Zone plate axis

Virtual image

Real image

Screen

phase. This aspect of holography plays a very important role in certain areas of its application.

To reproduce noncoplanar points and achieve depth, one replaces the multiple gratings with sinusoidal zone plates (see box). The algebra and geometry is much the same for both (one replaces plane waves with spherical waves), but the zone plate has one advantage over the grating, namely, it can image light to a point without a lens. Upon illumination by a collimated beam, it produces both real and virtual point images (again, see box). A hologram can be thought of as a series of overlapping sinusoidal zone plates that reproduce the points on the surface of the original object. Indeed, most holograms possess both real and virtual images, although the real image is more difficult to examine than the virtual image one usually sees. In our discussion, we will not use the sinusoidal zone plate description of a hologram, but give an alternate description based on a series of local sine gratings.

7.2 THE HOLOGRAM

With the preceding discussion of sine gratings as background, we can now generalize our explanation of the holographic principle. Consider the arrangement shown in Fig. 7.1, where a nonplanar reference wave is allowed to interfere with an arbitrary object wave at the holographic plate. If we consider the interference pattern over a region of the plate that is sufficiently small, we can approximate

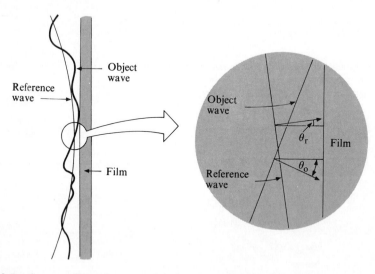

Fig. 7.12 In the recording of a general hologram, reference and object waves can be approximated by segments of plane waves over sufficiently small regions. These plane-wave segments interfere to produce local sine gratings.

the curved wavefronts by planar segments, as shown in Fig. 7.12. The angles the two waves make with the photographic plate, θ_o and θ_r, change from region to region. In each small region, however, the interference pattern produces a grating with a *local* frequency given by the expression

$$f_g = \frac{|\sin \theta_o - \sin \theta_r|}{\lambda} \tag{7.17}$$

which is a generalization of Eq. (7.6).

In examining the reconstruction of the hologram, we apply the same argument. In a small local region, the reference wave is approximated by a plane-wave segment making an angle θ_r with the hologram. But this can be shown to be the proper angle, according to Eq. (7.9), to reconstruct a plane-wave segment that exactly duplicates the object wave in that region. Each small patch of the hologram can be viewed as a local grating with the correct frequency to reconstruct a piece of the object wave when illuminated by the reference wave. Taken together, these pieces reconstruct the entire object wave. For good reconstruction, certain limitations should be observed. First, it is desirable for the reference wave to illuminate the hologram from a direction well separated from that of the object waves, so that the reconstructed object may be viewed without the eye intercepting light waves other than the desired object waves. Also, the magnitude of the reference wave should be constant across the holographic plate. If it is not, the local magnitude of the reconstructed object wave will depend upon the local magnitude of the reference wave and the original object wave will not be truly replicated. In regions where the reference-wave amplitude is too small, the object wave will reconstruct poorly, since the contrast of the recorded interference pattern is low.

Film Resolution, Source Coherence, and Stability

Holography involves the recording of an interference pattern on some light-sensitive material, usually photographic film (see box). Because of the short wavelength of light employed in optical holography, the fringe pattern to be recorded is usually microscopic in scale. As an example, consider the simple point source-point object arrangement of Fig. 7.8. Assuming $f = 10$ cm, $\lambda = 633$ nm, and $x = 2.0$ cm, the spatial frequency of the sinusoidal interference pattern that must be recorded on the film is approximately 300 linepairs, or dark-light cycles, per millimeter. In the simple three-dimensional hologram recording geometry illustrated in Fig. 7.1, local fringe patterns with spatial frequencies exceeding 1000 cycles per millimeter can occur. Such finely-structured patterns cannot be recorded with normal photographic materials (e.g., Eastman Kodak Plus-X film), which will record only spatial frequencies below about 100 lines/mm. The high resolving power generally required of holographic plates necessitates the use of special photographic emulsions, some of which are capable of resolving in excess of 2000 lines/mm. Eastman Kodak type 649F and Agfa Gevaert type 8E70 photographic

HOLOGRAPHIC RECORDING MATERIALS

A variety of materials other than high-resolution photographic plates and films can be used for recording holograms. Some of the more common are photoresist, thermoplastics, and photopolymers. *Photoresists*, commonly used in the manufacture of integrated electronic circuits, are photosensitive materials that form a surface relief pattern when properly exposed and processed. When the exposure is a holographic interference pattern, the resultant relief hologram can be used as a master for "pressing" soft plastic replica phase holograms* in a mass-production process. The principal disadvantage of photoresist materials is low sensitivity.

Thermoplastics are often used when the recording material must be used over and over in a near real time write-read-erase process. Usually a glass substrate is coated with a transparent electrode, a photoconductor, and the thermoplastic. The thermoplastic is then sprayed with a high-voltage electron charge and exposed to the holographic interference pattern. The key to the writing process is the photoconductor, which allows the originally uniform charge pattern to leak off in proportion to the exposure pattern. When the plastic is softened by heating (a current is passed through the transparent electrode), the remaining charge pattern causes a reliefing of the surface, with a phase hologram resulting. The hologram is erased by removing the charge and reheating.

Photopolymers are self-developing materials that can also be used in near real time holographic applications. They consist typically of a photosynthesized, incompletely polymerized polymer that becomes activated during exposure and polymerizes in the exposed regions during the succeeding several minutes, thereby forming a phase hologram. The materials are quite inexpensive. However, they are limited in their applicability by low sensitivity.

Other materials, such as photochromics and electrooptic crystals, are also used in holography, particularly in data storage applications. For a review of the common materials and an extensive list of additional references, the reader is referred to Ref. 7.9.

---

*See next subsection, Types of Holograms.

emulsions are examples. The price paid for this high resolving power is a very low film speed: These films must be exposed as much as 5000 times longer than everyday photographic materials to produce the same transmittance values.

The object wave and the reference wave exposing the holographic plate must be mutually coherent if their superposition is to result in the desired interference pattern. Hologram recording geometries should therefore be set up with attention given to equalizing object and reference beam pathlengths. If these pathlengths differ by more than the coherence length of the laser light, the object and reference waves overlap at the plate without forming an interference pattern.[†] It is helpful

---

[†] The coherence length for light from a laser depends on many parameters (the radiation linewidth for the laser line, the length of the resonant cavity, the presence or absence of an etalon, and so on). If it is not specified by the manufacturer, it can be estimated by simple experiment, as suggested in Section 2.6. See also Ref. 7.5, pp. 57–63.

to lay the recording geometry out on paper prior to setting up the optical system; a simple sketch will generally indicate whether or not nearly equal pathlengths can be achieved. At the same time, the recording geometry should be chosen such that the holographic image is physically separated from the undiffracted light (i.e., the transmitted beam) on reconstruction. The geometry shown in Fig. 7.1 has been arranged with such considerations in mind.

Even with the finest resolution film and a perfect recording geometry, one has no assurance of obtaining a good hologram. If the interference pattern exposing the photographic emulsion is to be satisfactorily recorded, the pattern must remain stationary during the exposure period. The movement of the pattern by even a small fraction of the spatial period of the fringes will reduce the recorded fringe contrast and hence reduce noticeably the brightness of the holographic image upon reconstruction. Fringe movements of the order of one period can prevent any image from being reconstructed. Since the positions or phases of the fringe patterns depend on the relative phases of object and reference light waves at the film, it is necessary that all components in the recording system that affect optical pathlengths–mirrors, beamsplitters, the object itself, and so on—remain fixed in position to within a small fraction of the wavelength of the exposing light during the exposure of the holographic plate. Even the air through which the light beams pass must remain relatively still, for variations in the refractive index of the air, caused by heat waves or turbulence, can produce a change in the optical pathlength in one arm of the recording system and a consequent "washing-out" of the recorded hologram. Except with special recording geometries or very short exposures, great care must be exercised in providing a vibration-free environment for the recording system. For this reason, hologram recording facilities are often highly specialized. Typical is a rigid, relatively massive tabletop supported on a tripod arrangement of pneumatic cylinders. The mass of the table provides mechanical inertia, and the pneumatic system effectively isolates the tabletop from room vibrations. Moving objects are often recorded with a pulsed ruby laser serving as the source of coherent light. The pulse duration can be adjusted from several milliseconds down to tens of nanoseconds, depending on the application. Holograms of exploding wires and high-velocity projectiles have been recorded using pulsed laser systems.

We do not mean to imply that holograms cannot be recorded with much less sophisticated equipment. Certain types of holograms can be recorded on an ordinary desk or table. A well-filled sandbox on small inner tubes can often provide the isolation from building vibrations required for most general hologram recordings. A number of practical guidelines for recording holograms and for testing system stability, coherence length, and film resolution are given in Ref. 7.5.

Types of Holograms

In our discussion thus far, we have made two simplifying assumptions regarding the nature of holograms. (1) The exposed and processed hologram modifies only the amplitude of the transmitted light in the reconstruction process and does not affect the optical phase of the light, and (2) the diffraction effects leading to the reconstructed image can be regarded as taking place in a very thin layer, essentially within a plane. These assumptions are valid for many, but not all, holograms.

The first distinction we can make between different classes of holograms is in the way in which light is diffracted by the hologram. Thus far, we have discussed only *absorption* holograms, wherein the *amplitude* of the incident light wave is modified in the reconstruction process through absorption by dark areas in the developed emulsion. A second type of hologram is the *phase* hologram, wherein it is the *optical phase* of the incident light that is modified in the reconstruction process, the amplitude of the light remaining unchanged. A phase hologram can be obtained by recording and developing a conventional absorption hologram and then bleaching it. In the bleaching process, the silver in the developed emulsion is replaced with some transparent salt of silver. This salt, silver bromide for example, has a refractive index different from the supporting gelatin matrix. As the incident light waves travel through the bleached emulsion, the local variations in refractive index introduce corrugations in the reference wave and, in the process, produce a replica of the original object wave. In the absence of Bragg selectivity, which we shall discuss shortly, a bleached hologram produces not only the desired image on reconstruction, but additional "ghost" images as well. Unless the recording geometry is properly chosen, these ghost images can overlap or interfere with the desired image. The principal advantage of the phase hologram over the absorption hologram is in diffraction efficiency, which is defined as the ratio of light power diffracted into the reconstructed holographic image to the power illuminating the hologram. The maximum diffraction efficiency for an absorption sine grating is 6.25 percent, and even this efficiency is not obtained in practice. For a thin sinusoidal phase grating, on the other hand, it is theoretically possible to diffract as much as 34 percent of the incident light into one of the diffracted beams. With the same illumination, a reconstructed phase hologram image is generally much brighter than its absorption hologram counterpart.

The second distinction we make between holograms relates to the thickness, or volume, of the holographic emulsion. Consider the diagram in Fig. 7.13. Plane waves interfere with spherical waves from source S in the region shown. The dark, curved lines show the locations of the interference maxima—the positions of greatest exposure for a photographic emulsion that intercepts the interfering waves. Holographic plates are shown in four different locations. Location ① is known as the *in-line* position and is the position of the holograms recorded by Gabor in 1948. Note that the fringes are widely spaced, suggesting the suitability of this location for recording holograms with low-resolution (and hence more sensitive) films. The disadvantage of the in-line location is that the reconstructed image is not separated from the undiffracted light and is consequently difficult to view.

Location ② is typical of hologram placement for *off-axis reference-wave holograms*, the kind we have discussed in earlier sections. For small angular offset (the average angle between the reference wave and the object waves) and for emulsions that are not too thick (several micrometers or less), the interference fringes exposing the hologram will have a separation large compared to their extent in the thickness of the emulsion. Holograms of this kind are referred to as *plane off-axis holograms*.

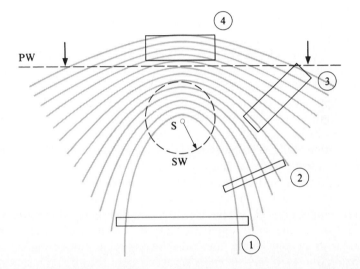

Fig. 7.13 Interfering plane wave (PW) and spherical wave (SW), represented by dashed lines, produce different grating characteristics in the holographic emulsion. Light lines show loci of interference maxima, where film exposure would be greatest. Thick and thin holographic emulsions are shown in four locations (see text): ① on-axis, ② off-axis reference-wave, ③ Bragg-effect, and ④ white-light.

As the angle of the reference wave increases to around 90°, as in position ③, the spacing between the exposing interference fringes becomes quite small. The silver remaining in the emulsion upon development of the hologram forms closely spaced and partially reflecting planes that diffract light on reconstruction only if the playback reference wave is incident on the hologram in a narrow range of angles about the original angle of incidence. For an emulsion 12–15 micrometers thick, for example, such a collimated reference-wave *Bragg-effect hologram**

---

* Named after W. L. Bragg, who first discussed such angular selectivity in connection with x-ray crystallography.

can be reconstructed only if the reference wave is incident at an angle within about 2° of the original recording angle.

When the holographic plate is exposed in position ④, the interference fringes are parallel or nearly parallel to the emulsion. Assuming the emulsion is sufficiently thick to record more than several dozen fringes in its volume, the processed emulsion will serve not only to reconstruct the hologram, but also as a selective interference filter, filtering out all but a narrow band of colors. Holograms recorded in this way are often referred to as *white-light holograms,** because they can be reconstructed with white light. The hologram itself, in filtering the light, introduces the necessary degree of temporal coherence, or monochromaticity. (The light source used to reconstruct the hologram must still exhibit good spatial coherence, and should therefore be compact, approximating a point source. Flashlights and streetlamps are often adequate; also the sun is often an excellent source.)

Another type of white-light hologram is the *multiplex hologram*, which can reproduce motion as well as depth. The construction process begins by filming with a motion picture camera the action to be shown in the hologram. It is necessary that there be a relatively circular motion of the camera about the subject. This is achieved either by moving the camera on a circular track or by placing the subject on a rotating table. The motion picture film is developed and a multiple hologram is made by imaging single frames of the film in sequence along with the reference wave onto a holographic film. To compensate for the distortion caused when the finished hologram is bent into a cylindrical shape, a cylindrical lens is inserted in the object beam path, compressing the image onto a 20 cm × 1 cm slit mask in front of the holographic film. The film is moved in small increments perpendicular to the slit length, and other overlapping holograms are made of the succeeding frames in the sequence. What has been recorded is a series of two-dimensional pictures which give information about the subject (1) from slightly different angles and (2) at different times. The developed and bleached hologram is bent into a cylinder and illuminated by a small line source along the cylinder axis. Because the relative motion of the camera has been translated into different angles on the holographic film, there is an illusion of depth, as there is with a conventional hologram. But, in addition, as the viewer walks past the hologram to observe new angles, additional frames are seen, giving the illusion of motion. Because of the nature of the construction process in which the real image is formed *at* the holographic film, the hologram may be constructed with white light despite the fact that the object/reference wave angle is less than 90°. Someday, when newer techniques have been developed, motion pictures based on holographic principles may be possible. For the present, Hollywood can only use holograms as props.

---

* Also known as Lippman-Bragg, reflection, or Denisyuk holograms, soviet physicist Y. N. Denisyuk being the inventor in 1962.

Before leaving our discussion of hologram types, it is appropriate that we consider briefly one final class of holograms, those generated by computer, or *computer-generated holograms*. If an object can be specified mathematically, it is possible for a computer to calculate the locations of interference fringes as they would be recorded on a photographic plate if this object were placed in a particular hologram recording setup. These fringes, or their equivalents, can be plotted with an $x–y$ plotter, and the resulting pattern photoreduced to a scale where it can be illuminated with the assumed reference wave. The object, which may never have actually existed, is "reconstructed" and can be examined from a number of perspectives. Computer-generated holograms can be recorded in several different ways. One large drawback is the great amount of computer processing time needed to generate the interference pattern. They have, nonetheless, an important role to play in holography, particularly in the area of optical data processing and in the testing of optical components.

7.3 APPLICATIONS OF HOLOGRAPHY

Holography has found application in a number of areas, and new uses are being developed all the time. The novelty and attractiveness of full-perspective, three-dimensional displays have inspired the limited use of holography in the advertising

---

SPECKLE INTERFEROMETRY

In addition to holographic interferometry, there are a number of related, nondestructive testing techniques, known collectively as *speckle interferometry*, that make use of the random laser speckle pattern discussed in Section 2.5. One such technique is *time average speckle interferometry*. The object to be tested, say a vibrating diaphragm, is made optically rough (with flat white paint, for example, to produce a good speckle pattern), illuminated with laser light, and photographed using high-contrast photographic film. At the film, the speckled image distribution is allowed to interfere with a reference wave that has been introduced by means of a beamsplitter. As the object vibrates, those regions of the image corresponding to the vibration nodes stand out as patterns of stationary, high-contrast speckle. In other regions, however, the speckles move about and become blurred due to the movement of the diaphragm. The bright, stationary speckles satisfactorily expose the film. The average irradiance of the smoothed-out portion of the speckle image is too low, however, to produce an adequate exposure. The result, on development, is a photograph of the object that shows clearly and with microscopic precision the regions of the diaphragm that were stationary and those that were moving.

Speckle interferometry, with its many variations, can be used with great versatility to measure object vibration, translation, strain, and rotation. Considering that speckle is often considered a nuisance in holography, it has clearly demonstrated redeeming features in other fields. The interested reader is referred to Ref. 7.10.

business. Of greater technical interest has been the use of holography in the high-accuracy measurement of object dimensions, deblurring of photographic images, and the characterization of low-density particulate matter. One especially interesting application has been in microscopy, where a hologram is made of an entire large-volume sample, freezing in one instant a record of many organisms suitable for later inspection at various depths under a microscope.

It is not possible to go into all the applications of holography, because the field is quite extensive and constantly growing. We shall look, however, at three important areas of application: (1) holographic interferometry, (2) holographic optical elements, and (3) holographic optical memories. Each of these applications exhibits a different aspect of holographic principles.

Holographic Interferometry

Holographic interferometry is a technique for measuring small displacements. (For an alternative technique, see box.) In one form or another, it can be used to observe and analyze, either in real time or after some delay, the microscopic flexure of a loaded support beam, to observe the strain in a fractured bolt, to record the shock wave from a bullet, to observe the vibrational modes of a kettle-drum head, or to detect hidden flaws in an aircraft tire.

The basic principles can be illustrated with the example of *single-exposure*, or *real-time*, *holographic interferometry* as applied to stress analysis. In this process, a conventional hologram is first made of the object to be tested. The hologram is developed in the normal manner and then returned to the exact position it held during exposure.* The object and the hologram are now illuminated just as they were during exposure, so that the reconstructed holographic image falls exactly on the object itself. If the object is now stressed slightly by heating or loading with weights, for example, fringes appear on it. These fringes result from the interference between the light waves from the now-distorted object and the holographically reconstructed waves from the "former" object. Where the distortion is small, the fringes are few in number; where it is great, many fringes appear. A direct qualitative indication of object stress is thus immediately available. An example of such a system of fringes appears in Fig. 7.14, where the object, a metal bar that is solidly attached at the bottom to a massive base, is stressed at the top. In the figure, the reilluminated bar and its holographic image are viewed simultaneously. Note that the rate of change of the beam displacement is greater near the top, the fringes being more closely packed there. Although in general the problem is quite difficult, with special geometries it is possible for one to interpret the fringe pattern to obtain a direct, quantitative measurement of object deformation (and, indirectly, of object stress). See Prob. 7.12, for example.

---

* An alternative is to develop the hologram in place. This can be done if the holographic plate is held in a special glass-sided container, which can be filled with the processing solutions.

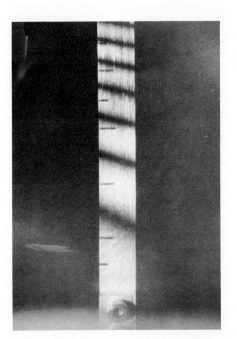

Fig. 7.14 Illustration of single-exposure holographic interferometry. Interference of the actual object wave and reconstructed object wave shows the amount of bending of the metal bar. The fact that the fringes are not horizontal indicates that besides the bend, the bar also suffered a slight twist about the vertical. The distance between marks on the bar is 1.27 cm (0.5″).

A variation of real-time holographic interferometry, known as *double-exposure holographic interferometry*, has found widespread application in several areas of nondestructive testing. Nondestructive testing of tires was, in fact, one of the first large-scale industrial applications proposed for holography. The basic approach is similar to the real-time technique just discussed. In the present case, however, the original and the distorted objects are both recorded holographically in a double-exposure hologram. When the hologram is reilluminated with the reference wave, both images are viewed simultaneously, and the interference pattern representing the differences in the two recorded light distributions appears localized on the object. If the object has changed shape during the two exposures, even microscopically, the change will manifest itself as a system of closely-spaced fringes in the reconstructed image.

A common application is illustrated in Fig. 7.15, which shows the double-exposure hologram of a portion of the interior of an aircraft tire. In preparation for the hologram recording, the tire is placed in a chamber. After the first exposure, the chamber is pumped down to a partial vacuum. The second exposure is then

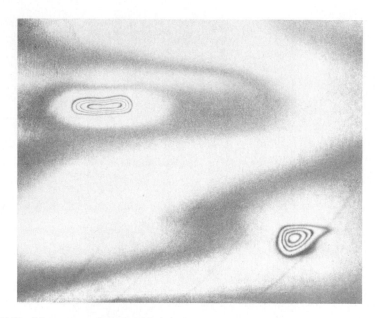

Fig. 7.15 Double-exposure holographic interferogram of an inside portion of an aircraft tire. Circular interference patterns indicate locations of tire ply separations. (Courtesy of Newport Research Corporation.)

made with the vacuum applied. If there is a local separation between the tread and the outer plies of the tire or between the various plies, both potentially dangerous flaws, air entrapped in the separation will cause a minute local expansion, a microscopic bulging, in the vicinity of the flaw. This bulging is readily discerned in the reconstructed hologram as a series of closely-spaced fringes. Commercial holographic testing systems developed for this purpose are able to test one tire every several minutes.

Closely related to single- and double-exposure holography is a technique known as *time-average holographic interferometry*. This technique allows the spatial characteristics of low-amplitude vibrations of an object—for example a drum head, speaker cone, or metal diaphram—to be mapped out with great accuracy. Figure 7.16 shows a vibrational mode of a guitar. The fringes represent contour lines of equal-amplitude vibrations of the guitar surface; the brightest contours occur at the nodes, or stationary points, of the vibrating surface. This is to be expected, because light reflected from these points always interferes with the reference wave at the hologram in the same way, yielding a recorded fringe pattern of high contrast and, hence, high reconstruction amplitude. For other

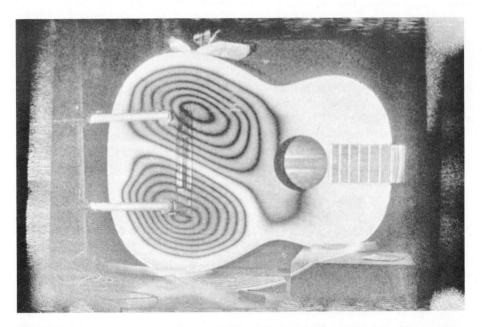

Fig. 7.16 Photograph of a time-average holographic image of a guitar. The fringe pattern shows the vibrational mode of the guitar at a particular frequency. (Courtesy of K. A. Stetson.)

regions of the vibrating surface, however, the corresponding interference pattern is moving and, for long exposures, produces a diffraction pattern of reduced or even zero reconstruction efficiency. The relationship between the pattern obtained and the vibrational amplitude at a particular location is not a simple one and is not derived here. It is possible, however, to use such time-average interferograms quantitatively.

Holographic Optical Elements

A hologram takes an incident light wave and, by means of diffraction, generates a different wave. In this sense the hologram can be viewed as a general kind of optical element with the capability of mapping one light distribution into another. The similarity between holograms and conventional optical elements has already been noted: A sine grating, the simplest of holographic records, changes the direction of a beam of light much as a prism does, and the sinusoidal zone plate behaves simultaneously like a positive lens and a negative lens.

This similarity between holographic elements and conventional optical elements can be exploited in a number of applications. Because of the strong wavelength dependency of the diffraction process (e.g., Eq. 7.6), holographic optical elements are highly dispersive; that is, they behave differently at different wave-lengths and must be used with monochromatic light or several of these elements must be combined in such a way as to eliminate wavelength-dependent effects. Nonetheless, holographic optical elements have a number of attractive features: They can be thin, lightweight, and compact; they can be easily fabricated and replicated; they can be very large in size. In addition, photographic emulsions can be coated onto arbitrary transparent surfaces for the fabrication of special-application, conformal optical elements. For example, one might manufacture holographic lenses that could be rolled up in tubes for storage or shipping.

One of the earliest and still most successful uses of a holographic optical element has been in optical spectrometers, where a high-resolution, holographically recorded grating replaces the considerably more expensive conventional grating. As holographic recording processes have been improved, so has the manufacture of these holographic gratings, to the point where holographic spectrometers now claim a substantial fraction of the spectrometer market. Similarly, holographic lenses (off-axis sinusoidal zone plates) have been used in imaging systems.

Just as glasses of different refractive indexes have been combined in the manu-facture of lenses that perform equally well over a wide range of wavelengths, so it is possible to combine multiple holographic zone plates and gratings in a system suitable for achromatic imaging over most of the visible spectrum. Although the individual component elements must be very precisely positioned for them to function properly, achromatic holographic imaging systems constructed in this manner exhibit the attractive features of small mass and reproducibility. To replicate such a system, it is only necessary to duplicate several holograms, whereas with conventional systems, glass elements must be ground and polished and tested.

Holographic optical elements are often used in optical data processing sys-tems. A particularly simple example is a coherent optical system (i.e., employing laser light) for recognizing alphanumeric characters. Let us consider in detail a subsystem for recognizing a specific letter—say the letter K. First, a hologram is made of a prototype letter K with type style and orientation appropriate to the recognition problem, as illustrated in Fig. 7.17(a). The reference wave is chosen to be a converging spherical wave. If the developed hologram is illuminated with the same spherical wave, the letter K is reconstructed. However, the converse is true also: If the hologram is reilluminated with light from the letter K, a spherical wave is produced that converges to a bright spot in the output plane of the optical system (Fig. 7.17b). There its presence can be detected with a photodetector. The light waves from other alphanumeric characters will also be diffracted by the hologram, but not into the suitably converging wave that is produced by waves from the correct letter (Fig. 7.17c). By detecting the presence or absence of a focused spot of light in the output plane of the system, one can determine whether a par-ticular character is present in the input plane. Additional holographic optical elements can be arranged to allow an entire page of printed material to serve as

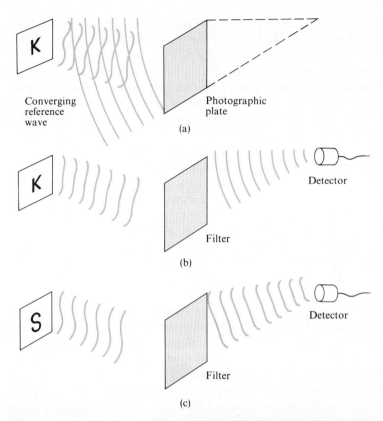

Fig. 7.17 Holographic character recognition (Vander Lugt filtering). The holographic filter, recorded as in (a), maps waves from a "matched" character into converging spherical waves, which are detected as points of light in the output plane (b). Waves from an incorrect character are also diffracted by the filter, but not into suitably converging waves (c).

input to the system. The locations of focused spots of light in the output plane then signify the locations of the corresponding "matched"* character in the input plane.

Holographic Optical Memories

The use of photosensitive materials in large-scale memories to replace magnetic-tape or disk bulk storage systems on computers has intrigued scientists and

---

* The system is the optical analog of the so-called *matched filter* receiver sometimes used in radar signal detection. The holographic optical element is often called a *Vander Lugt filter* after its inventor, A. B. Vander Lugt.

engineers for some time. Optical memories are limited ultimately in their information storage capacity by the wavelength of light used to "read" or "write" the information. Assuming an argon laser operating at 488 nm is the light source, optical memories are theoretically capable of storing in excess of 10^8 bits of information per square centimeter of photosensitive surface, or nearly 10^{13} bits in a 1-cm^3 volume. These numbers exceed the storage density limitations of magnetic systems by several orders of magnitude, and have made optical storage of digital information an area of extensive research. Much of the research conducted has centered on the difficult problem of preparing a photosensitive material with high resolution that can be written onto, read from, and erased with the fidelity of the recording process preserved over long periods of time.

The introduction of holographic techniques to the optical data storage area has led to optical memory systems for digital data that have several very attractive features. First, as noted earlier, holograms record information redundantly. Thus one bit of information, which in a conventional photographic recording process might be recorded as a transparent spot in a dark background, is in the holographic case recorded as a sinusoidal grating. The information for that one bit is spread out over the entire hologram. With conventional photographic recordings, a speck of dirt on the film could entirely block out the transparent spot corresponding to the bit. In the holographic case, a single speck of dirt has no significant effect on the recording. To recover the information recorded holographically, i.e., to "read out" the stored bit pattern, the holographic recording is placed in front of a positive lens and illuminated by collimated laser light. Each sinusoid generates a diffracted light beam, which is focused by the lens to a unique spot, there to be detected by a photodiode. The second advantage of the holographic recording of the bit pattern now becomes apparent, since the exact position of the hologram in the playback setup is seen to be unimportant. The hologram can be moved, either horizontally or vertically in the input plane, and the focused spots remain stationary. It is only necessary that the hologram be reasonably well illuminated. Binary information stored in sine-grating format thus exhibits simultaneously an immunity to "burst" errors (the speck of dirt) and an insensitivity to improper positioning in the readout system. These two characteristics have made holographic data storage attractive for moderately high-density storage of digital information when records are accessed frequently. Microfilm records of reports and journal articles are presently being made, for example, with cataloging information recorded in holographic form as well as in conventional form. Easy automatic readout of these data in a position-insensitive system is thus assured.

PROBLEMS

7.1   Referring to Eq. (7.1) describing the irradiance distribution resulting from the interference of two plane waves, under what circumstances does $I_o = I_m$?

7.2   Why must T_b and M in Eq. (7.4) satisfy the fundamental constraint $|T_b + M| \leqslant 1$?

7.3   The *diffraction efficiency* for a sine grating, η, is defined as the ratio of the light power in one of the diffracted beams to the power in the incident beam. In terms of the grating modulation constant, M, η for an absorption-type grating is given by $\eta = (M/2)^2$. Show that the maximum diffraction efficiency for a sine grating is 6.25 percent.

7.4   Following the approach leading to Eq. (2.24), show that Eq. (7.1) for the interference of two unequal-amplitude plane waves is correct. Assume that the two waves have amplitudes E_1 and E_2. (a) What is I_0 in terms of E_1 and E_2? (b) What is I_m in terms of E_1 and E_2?

7.5   For the case of normally incident plane-wave illumination, the angle of diffraction at a sine grating is given by $\sin \theta_d = \lambda f_g$. If the grating frequency, f_g, exceeds $1/\lambda$, then $\lambda f_g > 1$ and $\sin \theta_d > 1$, an impossible condition. Under these circumstances, no diffraction takes place. What is the maximum possible grating frequency, measured in grating cycles (or line pairs) per millimeter, if diffraction is to take place at a grating illuminated with the 514.5-nm light from an argon ion laser?

7.6   Referring to Fig. 7.4 for the $T-\varepsilon$ curve representing a typical holographic emulsion, let α be the slope of the curve in the linear region, i.e., $\alpha = dT/d\varepsilon$ in the vicinity of $T = T_b$. Assume that a grating is produced by exposing the film to the irradiance distribution of Eq. (7.1), and that I_0 and I_m are chosen to restrict $\varepsilon(x)$ to the linear portion of the $T-\varepsilon$ curve. What is M, the resultant grating modulation constant, in terms of α?

7.7   Figure 7.4 shows the $T-\varepsilon$ curve for a typical negative-working photographic emulsion (where increased exposure yields reduced transmittance). What would the corresponding $T-\varepsilon$ curve look like for a positive-working emulsion (where increased exposure yields increased transmittance)? Assuming that a suitable linear portion exists for a positive-working emulsion $T-\varepsilon$ curve, do you see any reason why such an emulsion could not be used in the making of sine gratings and, hence, holograms generally? Justify your answer.

7.8   A square wave, such as the one shown in the figure below, can be represented by a Fourier series of the form

$$f(x) = b_o + \sum_{n=1}^{\infty} b_n \cos 2\pi n f_0 x.$$

Show that a high-contrast diffraction grating that exhibits a square-wave transmittance profile, when illuminated with normally incident collimated light of wavelength λ, diffracts the light at multiple angles given by

$$\sin \theta_n = n\lambda f_g$$

where θ_n is the diffracted angle associated with the integer n, and f_g is now interpreted as the fundamental grating period. What is the maximum allowable value for the integer n for which diffraction occurs, given λ and f_g?

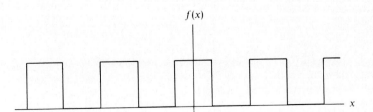

$f(x)$

x

7.9   Referring to Problem 7.8, discuss why high-contrast processing of holographic emulsions might lead to undesired "ghost" images in addition to the desired reconstructed images. What steps might be taken to eliminate these ghost images or at least guarantee that they do not interfere with the desired images?

7.10   The arrangement of mirrors, lenses, and beamsplitters used in recording a hologram must be carefully chosen. In the figure below is shown a laser, an object, and a photographic plate, all of which we assume must be held to these particular locations. Tracing these onto a sheet of paper, position the other components necessary to properly illuminate the plate and the object. Clearly identify each component and indicate the pathlengths traveled by object and reference beams. Be sure to minimize the pathlength difference.

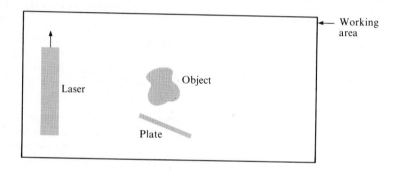

7.11   A one-milliwatt laser is used to reconstruct a $4'' \times 5''$ hologram. The beam is diverged by a microscope objective lens so that the beam overfills the plate by 20 percent (i.e., the beam cross-sectional area at the plate is approximately 25 square inches). If the plate attenuates the beam by 50 percent and 10 percent is lost to reflection from the plate surfaces, is it safe for a viewer of the hologram to view the transmitted reference beam directly (one microwatt/cm² at the viewer's eye is taken to be a safe level)? *Note:* If the irradiance at the eye is not at a safe level, a shadowbox should be built around the plate to protect the observer's eyes from direct illumination by the reference beam.

7.12   The displacement of a stressed beam attached at one end is proportional to the square of the distance along the beam from the point of attachment. Can you verify this relationship for Fig. 7.14? Plot the displacement in units of half-wavelengths as a function of distance along the bar. What is the displacement at the top of the figure if the hologram was made with a HeNe laser?

REFERENCES

Books

7.1   W. T. Cathey (1974), *Optical Information Processing and Holography*. New York: Wiley. Textbook written at the senior or first-year graduate level. Contains numerous problems and thorough mathematical development of holographic principles.

7.2 R. J. Collier, C. B. Burckhardt, and L. H. Lin (1971), *Optical Holography*. New York: Academic Press. The classic work on the theory and art of holography. Contains detailed descriptions and extended analyses.

7.3 J. B. DeVelis and G. O. Reynolds (1971), *Theory and Applications of Holography*. New York: Academic Press. Concise treatment of the subject at the graduate level.

7.4 J. W. Goodman (1968), *Introduction to Fourier Optics*. New York: McGraw-Hill. Textbook written at the senior or first-year graduate level. Develops the theory of coherent optical systems and holography in a modern linear-systems framework. Numerous problems.

7.5 M. Lehmann (1970) *Holography: Technique and Practice*. New York: Pitman (Focal Library Books). Monograph describing "how to do it." Not too heavy on theory.

7.6 G. W. Stroke (1969), *An Introduction to Coherent Optics and Holography* (2nd edition). New York: Academic Press. Coherent optics and holography presented at the graduate level. Contains reprints of Gabor's original papers on holography.

Articles

7.7 D. Gabor, W. E. Kock, and G. W. Stroke (1971), "Holography," *Science* **173**, 11–23. This and Ref. 7.8 are excellent review papers, written at an elementary level.

7.8 D. Gabor (1972), "Holography, 1948–1971," *Proceedings of the IEEE* **60**, 655–668.

7.9 R. L. Kurtz and R. B. Owen (1975), "Holographic Recording Materials—A Review," *Optical Engineering* **14**, 393–401.

7.10 K. A. Stetson (1975), "A Review of Speckle Photography and Interferometry," *Optical Engineering* **14**, 482–489.

8

Optical Communication

Since the first successful operation of a laser in 1960, optical communication has been one of the principle applications envisioned for this device. Communication by laser beam is attractive for several reasons. First is the extreme directionality of laser beams compared, for example, to the beams produced by typical microwave antennas. The directionality of a transmitted laser beam as determined by diffraction is expressed by the beam divergence angle, Φ (Section 2.3, Eq. 2.16):

$$\Phi \cong 1.27 \frac{\lambda_L}{D_L} \tag{8.1}$$

where λ_L is the wavelength of the beam and D_L the diameter of the beam waist, either in the resonator or at the output of a beam expander-collimator. This expression is also approximately true for the spreading of a pattern produced by a microwave antenna of diameter D_M radiating at wavelength λ_M. For the microwave transmitter to have the same beam divergence as a laser transmitter, however, the diameter of the microwave antenna must be

$$D_M = \frac{\lambda_M}{\lambda_L} D_L \cong 10^5 D_L \tag{8.2}$$

which is roughly 100,000 times that of the laser aperture. If the laser beam waist diameter is 1 mm, then the microwave antenna must be 100 m in diameter to achieve the same directionality.

Another reason that optical communication is attractive is the information-carrying potential of laser beams. The amount of information that can be sent over an electromagnetic wave is proportional to the bandwidth of the wave: 1000 times as much information can be conveyed each second in a 4-MHz bandwidth black-and-white television signal as in a 4-KHz telephone signal. As noted earlier, some lasers operate with exceedingly large bandwidths. For example, the bandwidth of a mode-locked Nd:glass laser that produces 30 psec pulses is approximately 1/(30 psec), or 30 GHz. With such a large bandwidth, it is theoretically possible to transmit five thousand color TV broadcasts over a single laser beam. This is to be compared with the four to five TV channels that can be transmitted simultaneously by a system with a bandwidth of 20 MHz (e.g., a microwave TV repeater system). The basic process of combining a number of different signals into one wide bandwidth signal is known as *multiplexing*. We will discuss several examples later in the chapter.*

While bandwidth and directionality are the two inherent advantages of laser communication systems, there are also several significant disadvantages. Although the laser possesses a large bandwidth, this bandwidth is not easily used. The usefulness of an optical system for communications purposes is generally limited by other factors such as detectors or the ease with which information can be imposed upon the beam. In addition, optical radiation is affected by turbulent atmospheric conditions and is significantly attenuated by rain or fog. Laser radiation does not reflect well from or penetrate through objects, such as trees or houses, as does a radio signal. These limitations, however, do not prevent the use of lasers in many special-purpose applications where other considerations may be of greater importance.

A specialized application that has received significant attention is the use of lasers in space communications, where atmospheric interference is not a problem, the distances are enormous, and the data rates and system weight are more significant than the cost of individual components. A second application is that of a rapidly installed, terrestial communications link for short distances, as between adjacent office buildings in a city.

8.1 THE LANTERN, THE HAND, AND THE EYE

One primitive method of communicating by light consists of holding a lantern in one hand and using the other hand as a shutter to vary the length of time that the light is visible to a distant observer. This system, although extremely simple, has all the basic features of a modern optical communication system. In the lantern, the moving hand, the air, and the observer's eye, we have the necessary

---

* The most common example of multiplexing in current telecommunications technology is the frequency-division multiplexing used in transmitting a large number of long-distance telephone conversations simultaneously over the same microwave communications link. The low-frequency voice signals, which have a range of frequencies from about 300 to 3500 Hz, are shifted in frequency, each individual signal occupying a different portion of the frequency band available to the microwave transmitter. One signal, for example, may occupy the range from 4 GHz to 4.000004 GHz (4 GHz + 4 KHz); another conversation occupies the range from 4.000004 to 4.000008 GHz, and so forth. In this way, thousands of telephone conversations can be transmitted simultaneously over the same microwave link.

light source, a device to impose a message on the beam, a medium through which the light is transmitted, and a detector, respectively. In this section, we discuss alternatives for each of these components. Because no other optical source approaches the laser in directionality, we restrict our attention to laser-based optical communication systems. Whether a gas laser, a doped-insulator laser, or a semi-conductor laser is the appropriate choice depends greatly on the application.

Modulation

The process of imposing a signal—be it a speech waveform, a TV signal, or binary data from a computer—on the laser beam is known as *modulation*. The device that performs the modulation is the *modulator*. Modulators in optical communication systems can range in complexity from a simple external, electromechanical shutter to specialized electrooptic and acoustooptic devices positioned within a laser cavity. One very simple way to modulate a laser is to vary the output power by changing the pump rate. Consider the example shown in Fig. 8.1, where the

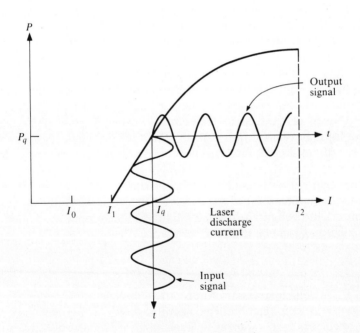

Fig. 8.1 Laser output power versus tube current for a HeNe laser. The input and output signals are variations about the operating (quiescent) point (I_q, P_q). For distortion-free signal output, the current variations must be restricted to the linear portion of the curve.

output power of a HeNe laser is plotted versus the laser tube discharge current. I_0 is the minimum current required to sustain the electrical discharge; I_1 is the threshold pump current required to initiate and sustain lasing.

If the tube discharge current is modulated in a two-level fashion between I_1 and I_2, the laser output is changed from full off to maximum output. Usually, variations in discharge current are restricted such that the laser operates in the linear region of the output-power-versus-current curve. There the change in output power due to a given change in power supply current, dP/dI, is approximately constant. This is a desirable attribute for many communication systems, since the variations in output power are then proportional to the modulating signal. As shown in the figure, if the input current from the power supply to the laser tube is varied sinusoidally about I_q, the output power of the laser also varies sinusoidally, with average value P_q. The point (I_q, P_q) is called the *operating point*, or *quiescent point*, of the laser.

In order to send a message over a laser beam, we must always modulate, or alter, the beam in some way. In all, there are five characteristics of a laser beam that can be changed: its power, frequency, phase, polarization, and direction. In the above example, the power of the laser has been modulated. In practice, direction and phase modulation are rarely used, and polarization modulation is used primarily as an intermediate step in modulating the beam power.

One of the more commonly used modulation devices is the electrooptic crystal (briefly discussed in the box in Section 5.3). If the polarization direction of linearly polarized light incident on a birefringent crystal is oriented at 45° to the fast axis of the crystal, the light can be considered to be split into two orthogonal components, one along the fast axis and one along the slow axis, each component traveling at its own speed determined by the refractive index for that polarization. The important feature of electrooptic crystals is that the difference between the refractive indices (the degree of birefringence) of the crystal can be controlled by an applied electric field. By varying the strength of the applied field, one can vary the amount that one orthogonal component is retarded in phase relative to the other. If an output polarizer, whose transmission axis is 90° to the input polarization, is introduced behind the crystal, the ratio of the output irradiance, I, to the input irradiance, I_0, can be shown to vary as

$$\frac{I}{I_o} = \sin^2\left(\frac{\pi}{2} \cdot \frac{V}{V_\pi}\right) \tag{8.3}$$

where V_π is the voltage necessary to produce complete transmission through the crystal—polarizer combination. This curve is shown in Fig. 8.2. Just as in the example of current modulation of a HeNe laser, it is often desirable to operate the system in the linear portion of the curve. Such operation can be simplified by applying a fixed dc bias voltage of value $V_b = V_\pi/2$ to the crystal.

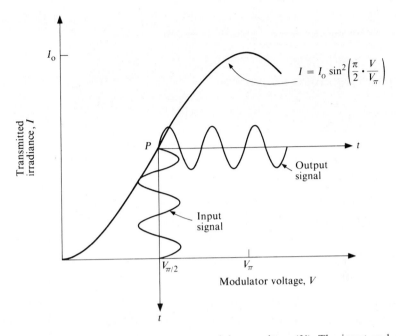

Fig. 8.2 Transmitted laser output (I) versus modulator voltage (V). The input and output signals are variations about a dc level.

Transmission

For transmission of an optical signal, one must decide whether to have the beam propagate through the atmosphere or to shield it with some kind of optical waveguide. Because of its simplicity, line-of-sight atmospheric transmission is appealing, but the signal degradation due to fog, rain, and atmospheric turbulence is too severe for this form of transmission to be attractive for systems requiring high reliability over propagation distances greater than several miles. Atmospheric path communication links between major metropolitan areas, for example, are generally not practical.

The alternative to an atmospheric propagation path is an optical waveguide. Thin cables of glass or plastic, called *optical fibers*, which confine the beam and yet are inexpensive to fabricate and maintain have been developed. An optical fiber, as illustrated in Fig. 8.3, consists of a thin, cylindrical core of glass of refractive index n_2 surrounded by another glass, called a *cladding*, of lower refractive index n_1. As shown, all rays of light incident from the core onto the cladding at an angle

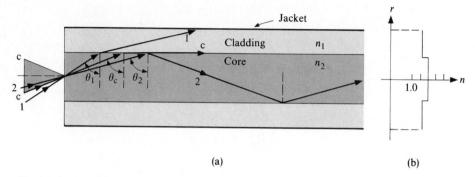

(a)   (b)

Fig. 8.3 Optical fiber. (a) Those rays in the central core having an angle of incidence θ_2 with the cladding greater than the critical angle, θ_c, are trapped and propagate down the fiber. Rays with an angle of incidence θ_1 less than θ_c pass through the cladding to be absorbed by the jacket. (b) Refractive index profile for the fiber.

with the normal greater than

$$\theta_c = \sin^{-1} \frac{n_1}{n_2} \tag{8.4}$$

are trapped in the core by total internal reflection. Rays of light transmitted into the cladding are absorbed by the jacket, which prevents crosstalk (interference) with other fibers. Many fibers can be combined in a bundle for carrying a great many signals simultaneously, and connectors exist now for splicing bundles together, much like telephone wires.

Standard fibers constructed of a separate core and cladding have the disadvantage that rays of light propagating at larger angles to the optic axis of the fiber require a greater propagation time than rays propagating at smaller angles. This effect leads to a broadening, called *differential delay*, of light pulses that propagate down the fiber. Differential delay can be virtually eliminated by fashioning fibers that have an extremely small refractive index difference between core and cladding, since the propagation is confined to very small angles with respect to the optic axis. As the ratio n_1/n_2 approaches unity, the critical angle approaches 90°, and only a single transverse mode (similar to the transverse modes of a laser, as discussed in Section 4.2) propagates. To ensure single-mode propagation at optical wavelengths, a refractive index difference of less than 1 percent and core diameters less than 5 μm are required.

An alternative to the standard clad optical fiber that greatly reduces differential delay while allowing multimode propagation is shown in Fig. 8.4. By properly doping the fiber during the manufacturing process, it is possible to produce a fiber that has a refractive index that decreases parabolically as a function of distance from the optic axis of the fiber. This refractive index gradient serves to

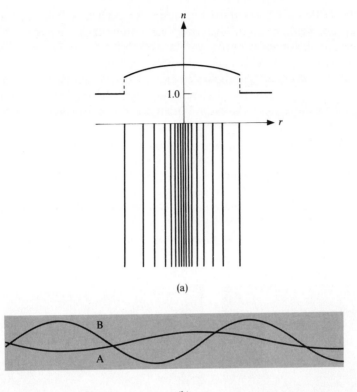

(a)

(b)

Fig. 8.4 Graded index optical fiber. (a) Fiber and refractive index profile. (b) Typical ray paths.

keep the optical beam confined to the center of the fiber, much like a distributed lens system.* Typical ray paths are shown in Fig. 8.4(b). Differential delay is greatly reduced, because the rays propagating at steeper angles to the fiber axis (*B* in Fig. 8.4b) travel in regions of smaller refractive index and therefore travel faster than the paraxial rays (*A* in Fig. 8.4b) do. So, despite the longer distance traveled, the nonparaxial rays arrive at the end of the fiber at the same time as the axial rays.

---

* In both the graded index optical fiber and the series of lenses, the optical path *nd* is smaller near the edge of these systems than it is in the center. In the lenses it is the thickness *d* that varies; in the graded index fiber it is *n*. The difficulty with using lenses is that they must be precisely aligned and closely spaced if a curved path is to be followed.

Optical fibers are particularly well-suited for use at near-infrared wavelengths. An important figure of merit for any optical communication system is the signal attenuation or loss over the propagation path. Fiber losses below 2 dB (approximately 37 percent) per kilometer have been reported.

Presently, two areas stand out as being especially important for optical fiber communication systems. The first is inner-city communication, particularly in dense population centers like New York City. Present telephone traffic in such areas is conveyed over complex mazes of wire pairs and coaxial cables, with so many wires and cables strung under city sidewalks and streets that there is little room for more. Fiberoptic links would require much less space to convey the same number of telephone conversations with the same reliability and, ultimately, at lower cost. Consider that a single optical fiber may be as small as 10 μm in diameter, that many thousands of fibers can be packed in a bundle the diameter of a pencil, and that with proper signal multiplexing each fiber can convey hundreds of telephone messages simultaneously. The potential is enormous.

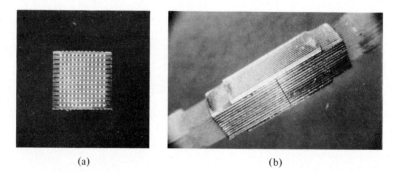

(a)   (b)

Fig. 8.5 Experimental fiberoptic splice. (a) End view of a 12 × 12 array of optical fibers held in place by grooved chips. The fibers and chips, set in epoxy, are highly polished. The array size is approximately 3 mm on a side. (b) Completed splice between two arrays. On the top and bottom are alignment chips that permit rapid and precise alignment of the arrays. Permanent splicing is achieved by applying index-matching epoxy to the assembly. (Reprinted with permission from the *Bell System Technical Journal*, Copyright 1975, American Telephone and Telegraph Company.)

There are many problems to be solved before fiberoptic systems can be used in everyday communication. In this technology, one cannot scrape the jacketing off the conductors and twist them together around a screw connector; totally new methods of connecting and splicing cables are needed. One experimental approach to splicing large numbers of optical fibers is shown in Fig. 8.5. Layers of optical fibers are arranged into a precise geometry by grooved spacer chips.

The array of fibers and chips is set in epoxy and the end of the bundle is polished smooth. The grooves on the outside spacer chips are used to align two polished arrays, which are then spliced with index-matching epoxy. The average loss across individual splices for this experimental array is 0.46 dB (10 percent).

The second area of special interest is in the transmission of communication and control signals on military vehicles, aircraft, and vessels. Conventional copper wire and coaxial cable-based signal links are subject to disruption by electric fields from nearby signal links or by the large pulses of electromagnetic energy that accompany nuclear explosions. Protection of vital electrical circuitry from such disruption is extremely difficult. Optical fiber, on the other hand, which is now replacing wire in many military installations, is virtually unaffected by electromagnetic interference. There is the added advantage, of course, of reduced weight and space required for what is usually many miles of communication links.

Detection

In any optical communication system, some form of detector is required to convert the variations in the received light back into variations in a signal voltage. Ultimately, whether it is the amplitude of the laser beam that is modulated or its frequency or polarization, it is incident irradiance to which the detector responds. Detectors used with lasers can be divided into two general categories: *thermal detectors*, which produce a response when heated by the absorption of radiation, and *quantum detectors*, which produce an electrical signal directly through the production of charge carriers (electrons or electron-hole pairs) by absorption of the light photons. With both thermal and quantum detectors, the response is proportional to the number of received photons. Since quantum detectors depend on the excitation of atoms to well-defined energy levels, the energy of the exciting photons must be greater than the gap energy ($hv > \Delta E$) for operation of the detector. Thermal detectors, on the other hand, respond to the total absorbed energy and do not exhibit a threshold wavelength for operation. Thermal detectors thus cover a wider range of wavelengths, but they are neither as fast nor as sensitive as quantum detectors. Fortunately, quantum detectors are available that are sensitive throughout the visible and near-infrared portions of the spectrum. Quantum detectors are employed in most optical communication systems. Thermal detectors are particularly useful for experiments involving measurements of high energies or for experiments at far-infrared wavelengths, for example, with CO_2 lasers. In this section, we describe two of the most common quantum detectors used in optical communication systems, the photomultiplier and the semiconductor photodiode.

Historically important, photomultipliers are still very widely used for the detection of optical radiation. They are especially appropriate when both high sensitivity and good response over a wide frequency bandwidth are required. The operation of the photomultiplier is based on the photoelectric effect. Light incident on a metallic surface, called a *photocathode*, can eject electrons from the surface, provided that the energy of each individual photon is sufficient to overcome the inherent binding energy of the electron to the surface. This characteristic binding energy is called the *work function* of the material. The lower the work function of the photocathode, the longer the wavelength of the radiation the photomultiplier can detect. Although a single photocathode can act as a detector, at low light levels the current produced is extremely small. Consider, for example, the detector response to a small helium-neon laser with an output power of 10^{-3} watts and a half-angle beam divergence of 10^{-3} radians. At a receiver distance of 10 km, the beam spot radius is approximately 10 meters. If the photomultiplier aperture is 1 centimeter in radius and the atmosphere attenuates the transmitted power by a factor of 100, then the received power is approximately

$$10^{-3} \text{ watts} \times 10^{-2} \times \left(\frac{10^{-2} \text{ m}}{10 \text{ m}} \right)^2 = 10^{-11} \text{ watts}$$

Assuming a photon detection efficiency of 10 percent for the photocathode, this incident power results in a current of

$$I = \text{number of electrons per second} \times \text{charge per electron}$$

$$= \frac{\text{laser power}}{\text{energy per photon}} \times \text{photocathode efficiency} \times \text{charge per electron}$$

$$= \frac{10^{-11}}{(6.6 \times 10^{-34}) \times (4.7 \times 10^{14})} \times 0.10 \times (1.6 \times 10^{-19})$$

$$= 5 \times 10^{-13} \text{ amperes}$$

This current is much too small to be conveniently processed and must consequently be amplified. Substantial photocurrent amplification is provided by the photomultiplier tube itself. The construction of a photomultiplier is shown schematically in Fig. 8.6. Each electron emitted by the photocathode is accelerated by an applied voltage to a secondary electrode, called a *dynode*. The dynode has the property of emitting several electrons for each electron that strikes it, thereby providing current amplification. These secondary electrons are then accelerated toward another dynode, where the process is repeated. Each dynode is held at a positive potential with respect to the dynode preceding it by means of a large voltage applied across a chain of resistors. A final anode collects all the electrons emitted from the last dynode. The current amplification achieved in typical photomultiplier tubes is on the order of 10^6. Incident powers as low as 10^{-19} watts can be detected using photomultipliers whose photocathodes have been cooled to reduce thermal noise.

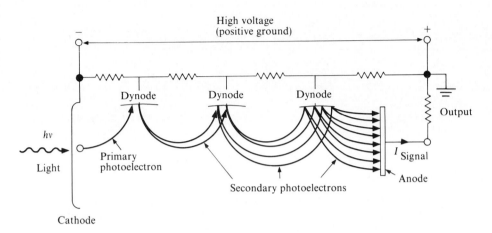

Fig. 8.6 Current amplification in a photomultiplier tube. Each dynode emits multiple secondary electrons for each electron that strikes it.

Photomultipliers are manufactured for use in a wide variety of circumstances. They are the appropriate choice of detector when light levels are low and bandwidths exceed several tens of megahertz.* Photomultipliers are generally bulky, but they are not especially rugged; they require supply voltages ranging from several hundred to several thousand volts. For these reasons, semiconductor photodiodes are often chosen as the detectors in optical communication systems.

Semiconductor photodiode detectors are widely used in the detection of optical radiation, not just in optical communication. Although they are not as well suited to low-light-level, wide-bandwidth detection as are photomultipliers, they are much less expensive, much more compact, and often require little more than a battery for a power supply. Because they are not particularly susceptible to damage by exposure to high irradiance levels, and because they have a reasonably rapid response time, semiconductor photodiodes are particularly useful in experiments involving pulsed lasers.

A typical photodiode consists of a section of p-type semiconductor material and a section of n-type material, forming a p–n junction, as described in Section 6.3. In operation, the junction of the photodiode is usually reverse-biased (the positive terminal of an external voltage source is connected to the n-type material), with the battery and photodiode connected in series with an external load resistor, as shown in Fig. 8.7. With no light falling on the photodiode, the junction is in equilibrium, and no current flows in the external circuit. When the junction is

---

* So-called crossed field multiplier phototubes have operational bandwidths that exceed one gigahertz.

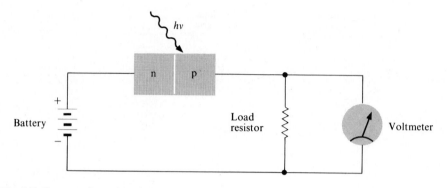

Fig. 8.7 Construction of a p–n junction photodiode detector. The photodiode is reverse-biased. With no light illuminating the junction, current in the external circuit is zero.

illuminated by light, charge carriers in the form of free electrons and holes are created, as illustrated in Fig. 8.8. The movement of the holes and electrons to establish the back-biased equilibrium creates a current in the circuit, providing an electrical signal to indicate that light has been detected. For high sensitivity, one can produce a device with a large junction volume by sandwiching an intrinsic semiconductor (doped to give equal numbers of holes and electrons, or undoped so that p = n) between the n-type and p-type regions. This p-intrinsic-n sandwich is called a *p-i-n photodiode.*

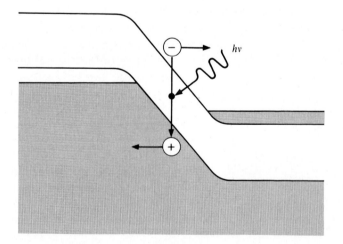

Fig. 8.8 Production of charge carriers by light. The light illuminates the center of the junction region, producing an electron-hole pair that contributes to a current in the external circuit.

A semiconductor photodiode is capable of photocurrent amplification, much like a photomultiplier tube, if the reverse-bias voltage is increased to near the breakdown voltage. Carriers moving through the junction region gain sufficient kinetic energy that they excite additional electrons from the valence band to the conduction band, thereby increasing the number of electron-hole pairs. Used in this mode, the device is appropriately called an *avalanche photodiode*.

8.2 DIRECT AND COHERENT DETECTION

As was noted earlier, not only the amplitude of a laser beam can be modulated to carry message-signal information, but also its phase, frequency, polarization, or direction. The type of modulation scheme chosen determines the nature of the detection operation that must be employed at the receiver, namely, whether it is *direct* or *coherent*.

Direct Detection

Direct detection is used whenever signal information is carried by changes in the irradiance of the received light (as opposed to changes in the optical frequency or relative phase of the light waves). The detection involves a simple conversion of the light variations into proportional variations in a signal voltage or current, which can then be processed by appropriate demodulation circuitry. Either thermal or quantum-type detectors can be used. Direct detection is often used in conjunction with *pulse code modulation* (PCM) of the laser beam. A typical PCM waveform is shown in Fig. 8.9(a). If the PCM signal waveform is thought to consist of a succession of time slots of duration T, the presence or absence of a pulse in any particular slot represents a "1" or a "0" in a binary sequence. A natural way of implementing this type of modulation scheme optically is with a mode-locked laser, which emits a regular train of pulses spaced by $2L/c$ seconds, L being the cavity length. In a process known as *pulse-gated binary modulation*, the output train of pulses from the laser is gated with an external modulator to produce a pulse train like the one shown in Fig. 8.9(b). Note that with this modulation technique, the duty cycle of the pulses (the ratio of pulse duration to pulse period) is very low, since the width of mode-locked pulses is short compared to the inter-pulse separation. The modulator, usually an electrooptic or acoustooptic device, must have a response time that is small compared to the interval between pulses, usually tens of nanoseconds. Note, however, that the modulator response time can still be long compared to the duration of the individual pulses. Often the receiver is gated on and off with the same period $2L/c$ in order to reduce the influence of noise on the detection operation. Many modifications of the basic scheme are possible. One alternative, illustrated in Fig. 8.9(c), is to represent "ones" and "zeros" by two orthogonal states of the laser beam polarization—either left and right circular or horizontal and vertical linear. A polarization-dependent beamsplitter, e.g., the birefringent crystal polarizer discussed in Section 2.7, then

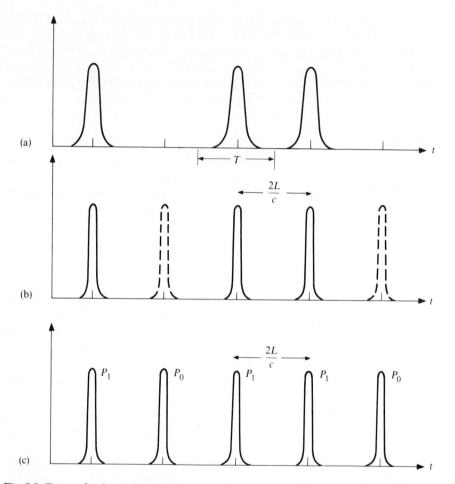

Fig. 8.9 Types of pulse code modulation. (a) Pulse code modulation waveform. Information is conveyed by the presence or absence of each pulse. (b) The same information is conveyed by pulse-gated binary modulation of a mode-locked laser. The pulse width is small compared to the interpulse separation. (c) The same information sequence conveyed by pulse polarization modulation of a mode-locked laser. Polarization states P_0 and P_1 are orthogonal, either right and left circular or orthogonal linear.

diverts the received beam to either of two detectors, depending on the polarization state. The principal advantage of this scheme, called *pulse polarization modulation*, over conventional pulse code modulation is that loss of signal is easily detected since the receiver expects a signal pulse at every time interval. There is the disadvantage, of course, that the receiver must have two channels, one for each polarization.

Coherent Detection

Most optical communication systems employ direct detection. It is possible, however, for signal information to be carried not by variations with time in the amplitude of the laser beam, but by variations in its phase or frequency. These variations cannot be measured directly: At the very high optical frequencies at which lasers operate, neither thermal nor quantum detectors respond to the instantaneous fluctuations in the incident electric field. They respond instead to the variations with time of the average power of the field. To measure variations in the phase or frequency of the incident waves, one must employ *coherent detection* methods. In essence, the phase or frequency of the received laser beam is compared with the phase or frequency of the waves from a stable reference laser, using interferometric techniques.

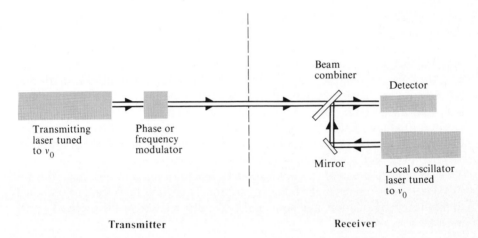

Fig. 8.10 Basic elements of a coherent optical detection system. Signal waves are combined with reference waves from the local oscillator laser, and the sum of the two waves is detected.

A schematic diagram for a coherent optical detection system is shown in Fig. 8.10. The signal-bearing laser beam is incident on a photodetector. Also incident on the detector is light from the reference laser, which is known as the local oscillator laser in analogy with the local oscillator used in the coherent detection of radio communication signals. Let $A_s \cos [\omega_s t + \varphi(t)]$ represent the electric field of the incident signal waves; ω_s is the *optical carrier frequency* (in the range 10^{13} to 10^{15} Hz, depending on the laser used), and $\varphi(t)$ represents the signal-bearing modulation. In the case of *phase modulation*, $\varphi(t)$ is proportional to the message signal; with *frequency modulation*, $d\varphi/dt$ is proportional to the message signal. The electric field of the local oscillator wave has the form $A_r \cos \omega_r t$, where ω_r is again very large, though not necessarily the same as ω_s. It is the superposition of these two wave fields at the detector that produces the total wave field,

$$E(t) = A_s \cos \left[\omega_s t + \varphi(t) \right] + A_r \cos \omega_r t \qquad (8.5)$$

The detector output voltage, $V(t)$, is proportional to the incident irradiance. Following the approach of Section 2.5 on interference, this can easily be shown to be proportional to $A_s^2 + A_r^2 + 2A_sA_r \cos \left[\omega_d t + \varphi(t) \right]$, where $\omega_d = \omega_s - \omega_r$. The optical frequencies, ω_s and ω_r, are much too high to be measured directly by the detector. The difference frequency (ω_d), however, can be chosen to be well within the response capabilities of the detector. If the local oscillator frequency equals the carrier frequency, then $\omega_d = 0$, and the detection operation is known as *homodyne detection*. If $\omega_r \neq \omega_s$, then $\omega_d \neq 0$, and the operation is known as *heterodyne detection*. In many optical heterodyne systems, ω_d is in the range of one to several hundred megahertz. The output signal is thus a phase or frequency modulated *radio frequency carrier* that can be converted back into the message signal with conventional phase-modulation or frequency-modulation demodulation circuitry.

As it is illustrated in Fig. 8.10, coherent optical detection is not generally a practical method for optical communication: Most laser oscillators simply do not have the frequency stability necessary to make this scheme workable. In analyzing coherent optical detection, we have assumed that the unmodulated transmitter laser and the receiver laser operate at *fixed* frequencies, ω_s and ω_r. We know, however, that laser light, although extremely narrowband, is not monochromatic. Frequencies ω_s and ω_r vary randomly in time, and instantaneous variations in $\varphi(t)$ cannot generally be distinguished from variations in ω_s or in ω_d ($= \omega_s - \omega_r$). An exception occurs with some longer-wavelength infrared lasers, including the CO_2 laser, as we discuss in the following section. Even here, however, one is limited to frequency modulation, and some noise-like degradation of the received signal must be accepted.

8.3 EXAMPLES OF OPTICAL COMMUNICATION SYSTEMS

A wide variety of laser-based optical communication schemes are possible, and many different combinations of sources, modulators, transmission channels, and detectors have been employed in actual systems. In this section, we describe briefly several different types of systems.

Fiberoptic Communication System

As suggested earlier, optical communication by fiberoptic transmission has been studied extensively, particularly by the Bell Telephone Laboratories, as an alternative to electrical communication via twisted pairs of wires and coaxial cables. The great advantage of the fiberoptic link is the extraordinary information-carrying capacity it has when compared to wire and cable links. A multifiber fiberoptic communication system is shown in Fig. 8.11. Semiconductor diode lasers (or light-emitting diodes) used in a pulsed mode serve as sources for the individual optical fibers of the system. Each channel has a bandwidth of at least several megahertz, allowing perhaps a hundred or more digitized speech signals to be multiplexed on a single fiber. Several thousand individual fibers may be contained in one transmission bundle.

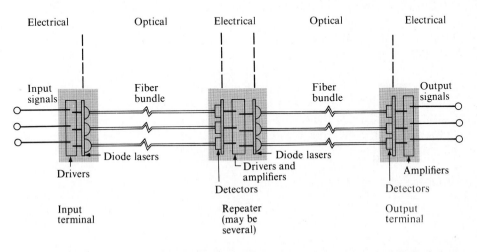

Fig. 8.11 Three-channel fiberoptic communication system. Each semiconductor diode laser represents one channel. Each channel has its separate optical fiber and associated repeater elements. The number of channels in a typical system could be several hundred or more.

Essential to the operation of a fiberoptic communication link exceeding several kilometers in length is a *repeater*, one for each channel, consisting of an input photodetector, a signal conditioner, and an output diode laser with driver. The signal conditioner takes the received pulses, which have been distorted by differential propagation delays, attenuation, and noise, and outputs a clean digital signal to the laser diode driver. Additional repeaters are added as the length of the link increases.

The optical receiver at the end of the link consists of photodiode detectors, conditioners, and appropriate demultiplexers and demodulation circuitry. The electrical signal-processing portions of the system are very similar to analogous portions of purely electrical digital-communication links. Discrete components can be used in the input and output sections of the system and in the repeater. However, considerable effort has gone into the development of *integrated optical circuits*, which perform the detection, conditioning, and retransmission operations all with a single small "chip." These integrated optical circuits are designed to include small optical waveguides, which allow the circuits to be coupled to the input and output optical fibers.

Coherent Detection Atmospheric Optical Link

The next system we consider is an atmospheric optical link using a CO_2 laser. The transmitted laser beam is frequency modulated; optical heterodyne detection is used to receive the signal. A simplified, functional block diagram for the transmitter is shown in Fig. 8.12. The transmitter laser is of the short cavity design and is operated single-mode and frequency-stabilized at one of the many 10.6 μm vibrational-rotational lines characteristic of the CO_2 laser. Tuning is achieved by changing the length of the laser cavity with a piezoelectrically actuated mirror. A stabilization system maintains the laser frequency at the center of the selected rotation-vibration line, usually the P(20) line at 10.5915 μm. Frequency modulation of the laser output is introduced by means of an electrooptic crystal (cadmium telluride) placed inside the optical cavity. A signal voltage applied to the crystal effectively changes the optical length of the laser cavity and thereby changes the single-mode resonance frequency. At the CO_2 laser wavelength, a change in cavity length by $\lambda/10$ changes the resonance frequency by 85 MHz, more than satisfactory for most FM communication systems.

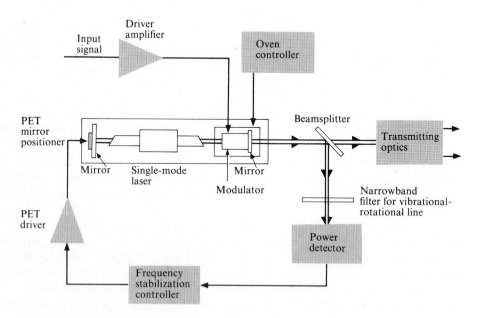

Fig. 8.12 Functional block diagram for a 10.6-μm optical FM transmitter system. Modulation is introduced by changing the cavity length and thus the output frequency of the laser.

The receiver, shown in Fig. 8.13, has a similar single-mode CO_2 laser serving as the local oscillator source. The received optical signal is mixed with light from the local oscillator laser at a mercury-cadmium telluride photodiode detector, which is cooled by liquid nitrogen. The output of this optical heterodyne detector, a radio frequency FM signal, is demodulated with a standard FM discriminator, which is a device that converts the fluctuations in instantaneous frequency of the detector output into changes in signal amplitude.

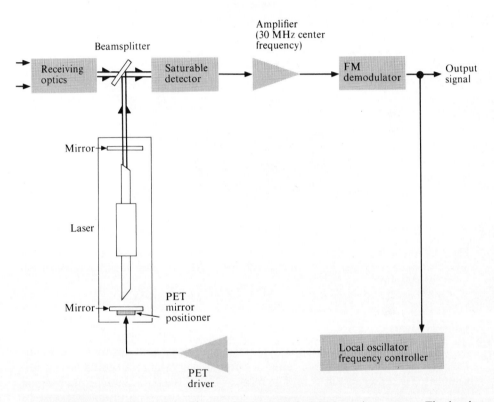

Fig. 8.13 Functional block diagram for a 10.6-μm optical FM receiver system. The local oscillator laser is tuned to a frequency 30 MHz above the mean frequency of the received signal laser beam.

Optical systems of this kind can be used for line-of-sight video (TV) signal links in all but the most severe weather conditions over distances exceeding 10 km. Because of their great directionality, laser communication systems of this kind are excellent for secure communication applications.

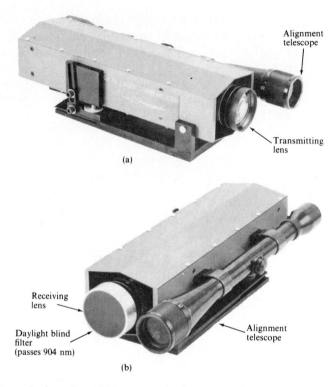

Fig. 8.14 Atmospheric path optical communication system for low data rates. (a) Gallium arsenide laser transmitter; (b) silicon avalanche photodiode receiver. (Courtesy of American Laser Systems, Inc.)

Other Systems

Atmospheric optical communication links are often built around a gallium-arsenide semiconductor diode laser. A commercial gallium arsenide laser communicator is shown in Fig. 8.14. The receiver uses a silicon avalanche photodiode with an optical bandpass filter centered at the laser's transmitting wavelength of 904.0 nm. Both transmitter and receiver use a 135 mm, f/2.8* lens to collimate and collect the light, respectively. The collimating lens reduces the large divergence of the semiconductor diode laser beam from about 150 mrad (full angle) to about 2 mrad. Although the transmission itself is entirely digital in nature (up to 10 Kbits/sec), both transmitter and receiver contain voice modulation or demodulation circuits, permitting use of either voice or digital data. For a two-way

---

* The f/number of a lens is the ratio of its focal length to its diameter. The irradiance at an image is inversely proportional to the square of the f/number of the imaging lens.

conversation, both users must have a transmitter and a receiver. The telescopes, mounted on the sides of the units, are used for visual alignment of the system. Peak laser power is 10 watts; pulsewidth is 100 nanoseconds. Such systems have been tested for ranges up to 24 km. Reliability exceeds 99 percent for ranges under 5 km in nearly all weather conditions.

The application of lasers to deep-space communications poses a variety of interesting problems. Of course, any type of closed pipe system is impossible in space, but atmospheric disturbance is nonexistent. Two principal types of systems have been considered: coherent systems using far-infrared, continuous-wave lasers and direct detection systems using mode-locked, solid-state lasers. An example of a solid-state laser transmitter that has been proposed for space applications is shown in Fig. 8.15. The laser is a mode-locked Nd:YAG laser operating at a wavelength of 1.06 μm. An external frequency-doubling crystal converts the near-infrared beam to the bright green at 0.53 μm, a region of the spectrum where detector sensitivity is high. The message is imposed on the mode-locked pulse train by an external modulator.

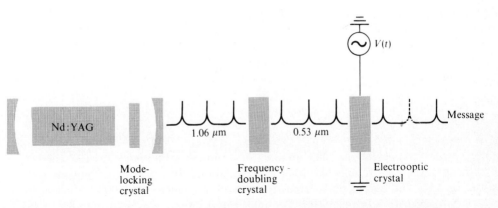

Fig. 8.15 Frequency-doubled, mode-locked Nd:YAG laser as a space communications transmitter.

The systems described thus far have not really exploited the tremendous signal bandwidth potential of laser communication systems, primarily because most optical communication systems built to date have been relatively conservative in their demands for laser bandwidth. The reason generally lies with the modulators employed: The laser itself may have an operational bandwidth of several hundred gigahertz, but more often than not, the range of frequencies over which the modulator can operate is limited to several hundred megahertz. One

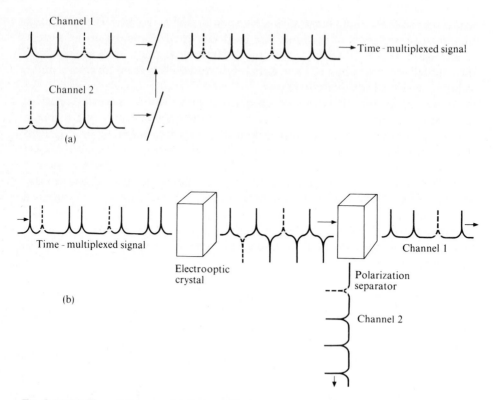

Fig. 8.16 (a) Time-division multiplexing of pulse-gated binary modulation signals. (b) Demultiplexing.

scheme for overcoming such modulator bandwidth limitations is *time-division multiplexing* of pulse-gated binary modulation optical signals, illustrated in Fig. 8.16. The output pulse trains from two pulse-gated, mode-locked lasers are combined with a beamsplitter. The interleaved pulse trains are separated at the receiver by passing the received light beam through an electrooptic crystal and a polarization-sensitive beamsplitter. The electrooptic crystal is driven so as to rotate the polarization of the second channel pulses by 90° with respect to the first. The beamsplitter then separates the two optical pulse trains, routing them to their respective detectors. More than two pulse-gated optical pulse trains can be multiplexed in this manner. The ultimate limit is determined by the ratio of the pulse repetition period to the duration of the individual pulses; we have shown this to be approximately N, the number of modes in the mode-locked output. Such a system might be useful for a high-data-rate, intercity link.

　　As noted in the introduction to this chapter, the invention of the laser was of immediate interest to telecommunications researchers due to the great potential of the information-carrying bandwidth of a laser beam. In spite of this initial excitement, however, optical communications technology has not yet progressed to the point where optical systems are replacing the conventional telecommunications systems in great numbers. Existing optical communications links convey only a very small fraction of the total telecommunications traffic of this country. In a sense, optical communications technology is still in its infancy. The basic techniques are well understood, but the manufacturing technology required to bring such systems into widespread application has yet to be developed. It is inevitable that such development will come, however, because the great potential that optical systems have for solving our expanding telecommunications needs still remains. The ultimate impact of optical communications technology will probably not come for some years, but when it comes, that impact can be expected to be enormous.

PROBLEMS

8.1 (a) Compare the divergence of a helium-neon laser with that of a microwave horn antenna transmitting at 30 GHz. Assume that the laser has a beam waist diameter of 1 mm and that the horn has a circular aperture diameter of 10 cm. (b) What is the irradiance, in watts/cm², at a receiver 1 kilometer away if the microwave antenna radiates 100 watts and the HeNe laser radiates 5 milliwatts?

8.2 Consider an optical fiber with refractive indices $n_{core} = 1.532$ and $n_{cladding} = 1.530$.

a) Calculate the critical angle for propagation.

b) For 1 km of optical fiber, calculate the difference in propagation time between a ray propagating along the optic axis and one propagating at the critical angle.

c) How does this compare to the duration of a mode-locked laser pulse?

d) Why is a small critical angle undersirable for most applications?

8.3 What is the critical angle for an optical fiber whose refractive indexes of core and cladding differ by 1 percent? by 0.01 percent? To what angle with respect to the fiber axis must the rays be confined to propagate in such fibers?

8.4 Calculate the detector current produced in a photodiode by a 1-μW helium-neon laser beam if each photon results in the creation of two charge carriers. What is the value of the load resistor that must be used for a voltage level of 10^{-3} volts?

8.5 Calculate the detector current produced by a 0.1-mW helium-neon laser beam incident on a photomultiplier consisting of a photocathode, five dynodes, and an anode. Assume that the photocathode has a quantum efficiency of 10 percent, that each dynode emits three electrons for each electron incident on it, and that the anode collects all electrons from the last dynode.

8.6 If the attenuation of an optical fiber, α, is given in decibels per kilometer (dB/km), the transmitted light power, as a function of distance traveled, is given by

$$I(z) = I_o\, 10^{(-\alpha z)/10}$$

where z is in kilometers. Given a one-watt signal at the input to a fiberoptic communication system, what is the maximum allowable distance to the first repeater if the attenuation is 10 dB/km and the minimum reliable signal level at the detector is 1 μW?

8.7   (a) Assuming that the air-fiber interface is cut square, as shown in the figure below, show that the "acceptance cone" angle θ of an optical fiber is given by

$$\theta_1 = \sin^{-1} \left\{ \frac{n_{\text{core}}}{n_{\text{air}}} \sqrt{1 - \left(\frac{n_{\text{cladding}}}{n_{\text{core}}} \right)^2} \right\}$$

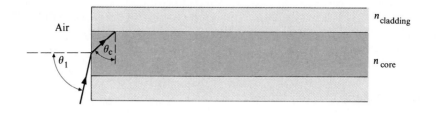

b)   For practical fibers, the indexes of refraction of core and cladding are approximately equal:

$$n_{\text{core}} - n_{\text{cladding}} = \delta, \qquad \delta \ll 1$$

Show that the expression obtained in part (a), assuming $n_{\text{air}} = 1$, reduces to:

$$\theta_1 = \sin^{-1} \sqrt{2\delta n_{\text{core}}}$$

c)   Calculate θ_1 for $n_{\text{core}} = 1.532$; $n_{\text{cladding}} = 1.530$.

8.8   Consider a CO_2 laser operating single-mode in an optical FM communication system like the one discussed in Section 8.3. Show that a change in optical cavity length of one wavelength (10.6 μm) will change the cavity resonance frequency by approximately 850 MHz.

8.9   (a) The full-angle beam divergence of a one-watt, ground-based laser transmitter is $\theta = 2 \times 10^{-3}$ radians. If the atmospheric attenuation is 10 db/km, what is the maximum allowable distance between repeaters if the repeater detector area is 1 cm² and the minimum reliable signal is 1 μW? (*Hint:* The detection may be limited by either the divergence *or* the attenuation.) (b) Using the same transmitter and receiver, what is the repeater distance in free space if the full-angle beam divergence is reduced to 1 μrad and the detector area increased to 10 cm²? (Assume that the attenuation of free space is zero.) Is the distance calculated on the order of satellite-to-ground, interplanetary, or interstellar distances?

8.10   (a) At approximately what rate can data be transmitted using a mode-locked argon-ion laser of cavity length $L = 1$ m as the basis of a pulse-gated binary-modulation optical communication system? Express your answer in Mbits/sec. (b) If a group of such systems is time-multiplexed to capacity, what is the data rate of the entire system? Assume that each mode-locked pulse is 50 psec long. (c) How many modes must oscillate to achieve the pulse length indicated in part (b)?

REFERENCES

8.1 S. S. Charschan (ed.) (1972), *Lasers in Industry*. New York: Van Nostrand Reinhold. Chapter 9, "Detection and Measurement," by S. S. Charschan is an excellent discussion of all aspects of optical detection. Many detailed illustrations are included.

8.2 M. Ross (ed.) (1971), *Laser Applications*. New York: Academic Press. Chapter 5, "Laser Communications," by M. Ross is a good overview of modern, high-data-rate laser communications. The author is an expert on pulsed lasers for space communications.

8.3 A. Yariv (1976), *Introduction to Optical Electronics* (2nd edition). New York: Holt, Rinehart and Winston.

8.4 A. Yariv (1975), *Quantum Electronics* (2nd edition). New York: Wiley. Chapter 14 discusses electrooptic modulators in detail.

8.5 W. S. Boyle (1977), "Light-Wave Communications," *Scientific American* **237**, 40. Additional information on the Bell System experimental optical fiber links.

9

Applications of
Laser Power

Perhaps more than any other attribute, the laser's capability to deliver large amounts of raw optical power or energy to a small region has brought it out of the research laboratory and onto the industrial production line. Only months after its development, the focused output of the ruby laser was used to pierce a hole in a diamond, the hardest material known. Five years later, a very short time as such things go, lasers were being used in the mass production of pierced diamond dies for drawing small-diameter wire. Presently, focused laser beams are used in materials processing applications in the automotive, aerospace, clothing, and pipe-line industries, to name only a few. From a user standpoint, materials processing is probably the best-developed field of laser applications and it promises to con-tinue growing. A related and exceedingly important potential application of focused laser beams is in the generation of power by laser-induced thermonuclear fusion. In terms of dollar volume, research and development in this area exceeds that in all other areas of laser application.

9.1 A POWER PRIMER

Before taking up these two areas of laser application, materials processing and laser fusion, it is appropriate that we consider in general terms those characteristics of laser light that make it so useful in such applications and the points that must be considered in the design of any laser-based system.

Materials processing is a field that uses the *power* output of a laser. Certain CO_2 lasers can deliver many tens of kilowatts of power on a continuous-wave basis, and peak powers from pulsed, solid-state lasers have approached 10^{12} watts. Although these tremendous amounts of power are available, the important parameter from a processing standpoint often is not simply total power, but *power density*, which is the number of watts of beam power delivered *per unit area*. It is because the coherent light from lasers can be focused so tightly with mirrors and lenses, resulting in enormous peak power densities, that the laser is so successful as a tool for cutting, melting, and otherwise working materials. As an added benefit, the weightless light beam is easily articulated with moving mirrors and lenses, and can be delivered in a highly controllable way to the material to be worked.

Powerful Lasers

Three particular lasers are used most often in laser power applications: the neodymium:YAG and neodymium:glass lasers and the carbon dioxide laser. Each of these has been discussed already in Chapter 6, but we consider them again as they relate to this special class of applications.

Although both the Nd:YAG and the Nd:glass lasers emit at the infrared wavelength of 1.06 μm, they are quite different in character. In one, the neodymium ion (Nd^{3+}) is doped into a host crystal, yttrium aluminum garnet (YAG). The Nd:YAG laser can be operated continuous-wave, but it is usually operated in a pulsed mode at rates up to 4000 pps (pulses per second). At a reduced pulse rate,[*] it can deliver as much as 10 joules per pulse, with 100 μsec pulse widths. Since the energy emitted in a Q-switched pulse must be stored in the medium (in this case, the crystal rod) prior to switching, the output energy of the Nd:YAG laser is limited by the small size of the synthetic YAG crystal, which can rarely be grown to a size larger than a pencil.

In the other neodymium laser, the ion is doped into a noncrystalline medium, glass. The storage volume of the glass laser can be considerably larger than that of the crystalline laser, because there are no size limitations on the medium beyond those imposed by uniformity of the material and the ability to cool it. The threshold level for the Nd:glass laser is somewhat higher than that of the Nd:YAG system, and the active volume is considerably larger. Consequently, large amounts of energy can be stored prior to Q-switching. One scheme for very large energy storage volumes is illustrated in Fig. 9.1, which shows a neodymium:glass laser amplifier made up of disks of neodymium-doped glass. Each disk is oriented at the Brewster angle to the incoming polarized beam to minimize reflective losses. The segmentation of the active medium allows efficient cooling, either by forced gas or forced liquid coolant flow.

---

[*] Allowing sufficient time for the laser material to cool.

Fig. 9.1 Neodymium:glass disk laser amplifier chain used in laser fusion research. Hinged flashlamp assemblies are exposed to view. Between the amplifier stages are Faraday rotators, which prevent amplification of back-reflected light. The disk areas are increased along the amplifier chain to keep the pulse energy per unit area below a maximum value to minimize nonlinear effects. (Courtesy of Lawrence Livermore Laboratory, University of California.)

When either of these Nd-doped lasers is simultaneously Q-switched and mode-locked, extremely short pulses of picosecond (10^{-12} sec) duration and extremely large peak powers are obtained. The glass medium broadens the neodymium ion lasing transition more than the garnet crystal does; therefore, as was discussed in Section 5.4, more modes are available for mode locking and sharper pulses are produced. In some systems, such as the one-kilojoule system discussed later, a Nd:YAG oscillator is used in combination with Nd:glass amplifiers.

The other laser frequently used for power applications, the CO_2 laser, can furnish up to tens of kilowatts of power when used continuous-wave. TEA configurations, which are generally pulsed, can provide several hundred joules in a 100-nsec pulse, corresponding to peak powers of approximately a gigawatt (10^9 watts). Because its output wavelength is at 10.6 μm, the CO_2 laser is especially useful for processing rubber, plastics, glass, wood, cloth, and ceramics, all of which absorb well in the infrared. Even metals, which are reflective at 10.6 μm, will break down in the intense focused beam of the CO_2 laser.

Industrial CO_2 lasers are usually constructed in a flowing-gas configuration, as opposed to the sealed-tube configuration characteristic of their low-power relatives. Figure 9.2 shows the exterior of a 10 kW flowing-gas system built for research into industrial applications. In the design of such a high-power system, great care must be exercised in the specification of optical components. Mirrors are often made of copper, which is highly reflective at 10 μm. These are cooled by heat pipes and forced-flow coolant that remove the fraction of the beam energy that is absorbed. For CO_2 lasers having a lower power output (~ 1 kW CW), focusing lenses of gallium arsenide or germanium are used, since these two substances transmit well at 10.6 μm.

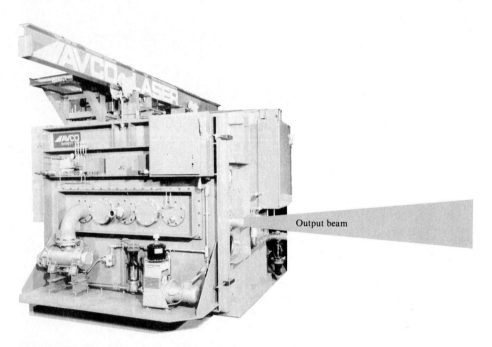

Fig. 9.2 Photograph of the main lasing chamber of a 10-kW industrial CO_2 laser. The chamber includes an electron beam source for pumping the CO_2 gas, gas circulation system, and heat exchanger. (Courtesy of Avco-Everett Research Laboratory, Inc.)

Focusing Laser Beams

The 50-watt output of a small industrial CO_2 laser does not seem like much compared to the 1000-watt rating of a tungsten filament heat lamp. But, as we noted earlier, it is the ability of the laser to deliver high power densities to a small area that makes the laser attractive for many applications. This is because the high degree of transverse coherence permits the light from a laser to be focused onto a much smaller area than can the light from a more conventional source. Still, diffraction-spreading imposes a minimum on the achievable diameter of the focused spot. Using diffraction theory, one can show that the Gaussian beam from a laser operating in the TEM_{00} transverse mode can be focused to a spot of radius

$$w_0 = \frac{\lambda}{\pi} \cdot \frac{f}{w_1} \tag{9.1}$$

where f is the focal length of the focusing lens used, and w_1 is the spot size, or radius, of the (collimated) beam incident on the lens, as measured from the beam center to the point where the field amplitude is down by $1/e$ (the irradiance is $1/e^2$ of that at the beam center), as shown in Fig. 9.3. Equation (9.1) assumes that the beam is significantly smaller than the lens itself and that the diffraction spreading is a consequence of the Gaussian beam profile and not of any aperture in the focusing system. In certain applications, the collimated beam may "overfill" the lens or associated aperture. In such cases, the spreading is described by the diffraction at a circular aperture, discussed in Section 2.3. If the lens or aperture is nearly uniformly illuminated, the radius of the focused spot of light is

$$w_0 = 1.22 \frac{\lambda}{D} \cdot f \tag{9.2}$$

where w_0 now defines the radius from the center of the spot to the first dark ring of the diffraction pattern, and D is the diameter of the aperture. Note that in both Eqs. (9.1) and (9.2), the focused spot size is proportional to the wavelength. For this reason, neodymium lasers are sometimes used in applications where a CO_2 laser might otherwise serve were it not for particularly stringent focussing requirements.

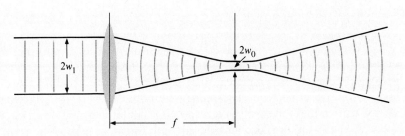

Fig. 9.3 Focusing of a collimated Gaussian (TEM_{00}) laser beam of spot size w_1, with a spherical lens. The spot size in the focal region is $w_0 = \lambda f / \pi w_1$.

Related to the spot size of the focused beam is the distance, in the region of the focused spot, over which the beam diameter remains small. This distance, known as the *depth of focus*, is important in a materials processing operation, for it determines how carefully the workpiece must be positioned within the focus, or how deep the beam will cut without refocusing. The depth of focus is somewhat arbitrary in that it depends on the beam focus tolerences one chooses to apply. For the Gaussian beam considered above, a useful definition for depth of focus has the form

$$d = \frac{2\lambda}{\pi} \sqrt{\rho^2 - 1} \left(\frac{f}{w_1}\right)^2 \qquad (9.3)$$

In this equation, ρ is the tolerance factor: If within the distance d the spot size is to increase by no more than 10 percent (0.10) from its minimum value at the waist, then $\rho = 1.10$. Expressed in terms of the minimum spot size, w_0,

$$d = \frac{2\pi}{\lambda} \sqrt{\rho^2 - 1} \, w_0^2 \qquad (9.4)$$

Note that a large depth of focus is not compatible with a small spot size.

As an example, consider a CO_2 laser with collimated beam radius $w_1 = 4$ mm. A germanium lens with a focal length of 60 mm can focus the 10.6-μm beam to a spot with a radius

$$w_0 = \frac{10.6 \times 10^{-6} \text{ m} \times 60 \times 10^{-3} \text{ m}}{\pi \times 4 \times 10^{-3} \text{ m}} = 50 \ \mu\text{m}.$$

The depth of focus for a 5-percent allowable increase in spot size is

$$d = \frac{2 \times 10.6 \times 10^{-6} \text{ m}}{\pi} (1.05^2 - 1)^{1/2} \left(\frac{60 \times 10^{-3}}{4 \times 10^{-3}}\right)^2$$

$$= 0.49 \text{ mm.}$$

With a 250-mm lens, the depth of focus is extended to approximately 8.5 mm, but the minimum spot size also increases to more than 200 μm.

When so much energy is focused into such a small area, enormous electromagnetic fields result. For example, a pulsed laser with a peak power of 10 megawatts focused into an area 30×10^{-6} cm^2 (corresponding to a beam radius of 30 μm) produces an electromagnetic power density of approximately 3×10^{11} W/cm^2 at the focus, with a corresponding field strength of 1.5×10^7 V/cm, a value well above the breakdown field strength for most gases. The result is a spark, similar to that at the tip of a sparkplug or in a lightning flash. When impurities such as material vapors from laser processing are present, the field strength necessary for breakdown is even smaller than that for a high-purity gas. Such breakdown is to be avoided in materials processing, since any energy that has been expended in ionization of the atmosphere is unavailable for machining purposes. Therefore, if there is a need to obtain power densities beyond the point of gas ionization, the light pulses must be transmitted in a vacuum.

Material Properties

The successful application of the laser depends not only on selecting the appropriate laser and beam focus geometry, but also on understanding the properties of the materials to be processed. Probably the most useful set of data for processing applications is the dependence of the absorptivity and reflectivity of the material as a function of wavelength. Obviously, if a material does not absorb the energy incident upon its surface, it will not heat to the melting or vaporization point. An

---

FOCUSED BEAMS FOR LOW POWER APPLICATIONS

Although the emphasis in this chapter is on the use of focused high-power laser beams, it is appropriate to mention two low-power applications which represent the first laser-based devices to be used by the public at large. The first of these, the *point of sale device* (POSD), is used to price items in supermarkets. The key to this system is a machine-readable label printed on most products by the manufacturer. The label is a series of parallel bars of varying width encoding numerical information known as the Universal Product Code (UPC). In operation, the device scans a focused laser beam back and forth across the label many times each second in an intricate design. A detector, sensitive only to the laser wavelength, detects the variations in light reflected from the dark and light bands of the label and converts them to electrical signals. When the label is scanned, the price of the identified item, posted at the shelf and stored in the computer, is added to the customer's bill. Besides eliminating manpower costs for marking each item in the store, the POSD can decrease checkout times and provide up-to-date inventory control.

The second application introduces the laser into the home as a space-age version of a phonograph needle for a *videodisc system*. The videodisc is similar to a phonograph record, but contains video information that can be converted to a standard video signal for playback on the home television set. Because the rate of information that must be transmitted is so much greater than that for an audiodisc, the rate of rotation and the density of information on the videodisc must be greatly increased over the values for a phonograph record. A readout using a focused laser beam holds several advantages. Since there is no contact with the disc surface, there is no wear due to playback as there is with records. The density of information is easily accommodated by the tightly focused beam. Furthermore, it is easy to "freeze" the picture by maintaining the beam on the same track. This enables catalogs and printed material, as well as standard television fare, to be stored and read back through the home television set.

absorption-versus-wavelength curve for a material is therefore an important aid in determining the proper laser. The thermal conductivity and the heat capacity of a material are also properties to consider in processing. If the material is too thick or the thermal conductivity too high, the process may not achieve the desired result. Information on most materials is available in handbooks and applications notes from laser manufacturers. When no such data exist, the material itself must be subjected to a number of tests with different lasers to obtain the necessary information.

A Calculation

A simple calculation serves to illustrate the quantities that must be considered in a typical materials processing application. Assume we want to cut a thin sheet of iron alloy having the following properties:

Thickness:	2 mm
Density:	8 g/cc
Specific heat:	0.125 cal/g (1 cal = 4.2 joules)
Reflectivity:	95% at 1 μm

We use a Nd:glass laser with the following characteristics:

Wavelength:	1.06 μm
Pulse energy:	4 joules
Pulsewidth:	1 μsec
Output beam radius:	0.6 mm

When this output is focused by a 53-mm focal length lens, using Eqs. (9.1) and (9.3), the spot size and the depth of focus ($\rho = \sqrt{2}$) are 30 μm and 5 mm, respectively. Assuming that at least 5 percent of the laser energy from the Nd:glass laser is absorbed (95 percent reflected) in a cylindrical volume 5.65 × 10^{-6} cm^3 (radius = w_0; height = sheet thickness), what is the maximum temperature rise in this volume? Multiplying the specific heat of the material by this volume and by the density of the material, the amount of thermal energy needed for a one-degree temperature increase is found to be 24 × 10^{-6} J/°C. Dividing this into the heat absorbed (0.2 J), a temperature rise of about 8400°C would be indicated. The heat generated is not confined entirely to the focus volume, but is rapidly conducted away. Still, the temperature rise is substantial, in many cases above the melting point of most metal alloys (~ 1500°C), depending upon the power of the laser beam. If the temperature rise occurs rapidly, the accompanying thermal expansion of the heated zone in the vicinity of cooler material introduces enormous strains in the sample. These strains are relieved by the propagation of high-intensity compressional waves away from the laser-damaged area. In some cases, a miniature explosion of material from the focus point may result. It is the effects of these gigantic temperatures and pressures that we want to discuss now.

9.2 MATERIALS PROCESSING

Lasers have found numerous uses in materials-processing applications where great precision and controllability are important system characteristics. The laser beam is easily manipulated; the mirror systems used in directing the beam can respond quickly and easily to control commands. Except when delivered in extremely high-power, short-duration pulses, the laser beam imparts virtually no mechanical stress to the workpiece. The laser beam is highly versatile, capable of tempering, melting, cutting, or scribing the material in the beam focus, depending on the amount of beam energy and the manner in which it is delivered.

A Simple Example

One example of using the laser to advantage in manufacturing is to puncture baby-bottle nipples. Three holes must be put in the nipple, one at the top of the nipple for the milk outlet and two air inlets at the base of the nipple. Formerly these holes were made by puncturing the rubber with hot wires, which accumulated a gooey residue after a period of time and had to be cleaned. This hot-wire method can be replaced by a laser-beam perforation method. The laser output, split into three beams, is directed to the appropriate points on the precisely positioned baby-bottle nipple. The laser is pulsed, the required holes created, and the next piece rotated into proper position.

There are a number of points that should be considered in designing a materials processing system—they are listed in Table 9.1. To illustrate some of these points, we follow the above example a little further. The type of laser chosen must have an output wavelength at which the rubber is absorptive. A laser whose output is highly reflected by the rubber or the charred surface of the rubber is inefficient or even useless. If there is a choice to be made between a continuous-wave and a pulsed version of the laser, a pulsed laser might be better suited to this hole-forming application, since pulse powers can be quite high and no purpose is served by using a CW laser in a repetitive process such as this. As a rule, it is less expensive and less complicated to build a pulsed laser than its CW counterpart, because both the power and cooling systems can be considerably simpler for the pulsed version. The number of pulses and their duration are other factors that must be considered. Whether a single pulse or a series of repetitive pulses is needed to create the hole in the rubber depends upon the energy in the pulse and the manner in which the rubber is destroyed. There may be advantages to Q-switching the laser to obtain a high-power pulse and then providing time for the vaporized gases to be dispersed before initiating another pulse.

Table 9.1 Design considerations in materials processing systems.

1. Source
   a) Wavelength
   b) Power level
   c) Pulsed or continuous-wave

2. Beam handling
   a) Fixed or scanned
   b) Focus lens
   c) Gas jet assistance

3. Materials handling
   a) Piece positioning
   b) Processing speed
   c) Monitoring for completion of the process

4. Safety
   a) Beam security
   b) Workpiece and processing area enclosure
   c) Safety monitoring
   d) Personnel locations during processing

If previous data exist on the dimensions of the volume of material to be removed, as there certainly does in our example of the baby-bottle nipples, the focal length of the focusing lens must be chosen to give the appropriate beam spot size and depth of focus to match the hole size requirements. The stable deflection and positioning system for the laser beam should be well thought out, as should the design of the device for positioning the material to be processed. Finally, the beam or some portion of it should be used to monitor the progress of the operation and to provide safety indicators for the duration of the process. Enclosures should be designed to protect against errant laser beams and stray reflections. No laser can be considered truly useful unless it is safe for the operator(s) and workers in the vicinity of the laser.

Heat Treating

Metals and certain other materials are often heated for a period of time to harden them. Such heat treatment is common in the tooling and automotive industry. In certain cases, it is desirable to treat only a portion of a larger item. For example, an automobile piston must be hardened in manufacturing, yet heat treatment of the entire piston could lead to undesirable dimensional changes. This is an excellent example of a situation where the ability to focus the laser beam precisely makes it particularly advantageous in materials processing. As shown in Fig. 9.4, the beam

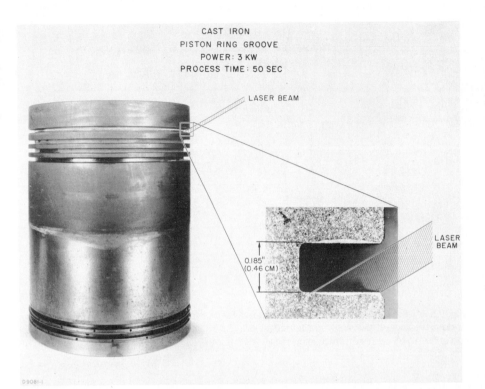

CAST IRON
PISTON RING GROOVE
POWER: 3 KW
PROCESS TIME: 50 SEC

LASER BEAM

LASER
BEAM

0.185"
(0.46 CM)

Fig. 9.4 Laser heat-treating a ring groove of a cast iron piston. The depth of the treated area is indicated in the inset of a cut section of the piston by the light region at the top and bottom of the groove. No detectable distortion results from this laser process because heat is delivered only to the region to be hardened. (Courtesy of Avco-Everett Research Laboratory, Inc.)

can be directed to affect only one surface of the ring groove. The facing surface can be treated by repositioning the beam onto a mirror that directs the beam upward.

The output beam from the laser system that is used to process the automobile piston has an annular cross section. This output beam shape lends itself to a clever solution to heat-treating a metal shaft. The beam is focused with a toric mirror encircling the shaft, allowing the entire circumference to be treated simultaneously. In Fig. 9.5, a cross section of a treated shaft is shown along with a plot of metal hardness as a function of the distance from the surface. We see from the plot that a small, well-defined region at the surface of the shaft has been hardened.

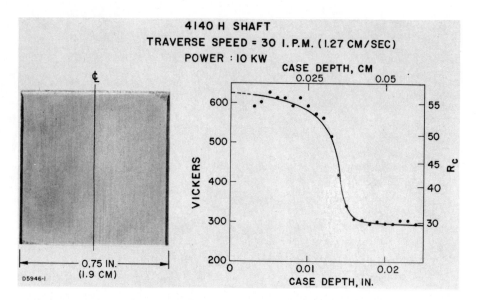

Fig. 9.5 Cross section of a shaft heat-treated by the laser system shown in Fig. 9.2. The darkened areas on the right and left edges are the hardened regions. The graph is a plot of metal hardness versus distance from the surface. (Courtesy of Avco-Everett Research Laboratory, Inc.)

In some cases, the shape of the part is so intricate that a simple focusing geometry cannot be used. In that case, mirrors are used to scan the laser back and forth over the region to be treated. Alternatively, the region can be painted with absorbing material before processing to increase the local absorption of laser radiation.*

If the power of the laser beam is increased somewhat over that used for pure heat-treating, a thin layer of melted metal can be formed on the surface. When the proper elements are introduced into this molten layer, an alloy entirely different from the bulk metal is formed. This alloying process enables one to cast a part in a cheap base metal and then selectively harden the areas of greatest wear.

Welding

With increased power output, it is possible to use the laser (primarily CO_2) as a welding tool. Because the affected area is small, the work can proceed rapidly and with a minimum amount of distortion of the surrounding area. Welding rates

---

* This same principle has been used on priceless sculpture to remove a black encrustation deposited by air pollution. The laser cleaning method is preferred to other methods because it does not abrade the surface (as does sandblasting), and yet it removes debris from deep grooves in the surface of the sculpture.

for a type of stainless steel (304) can be as high as 10 cm/sec for a 5-mm penetration or as low as 1 cm/sec for a 12-mm deep weld using a 10-kW CO_2 laser. An extremely complex application is the welding of underbodies of automobiles. By computer controlling the beam deflection mirrors, it is possible to weld a curved contour while the entire piece is moving at 1.5 m/min.

An application that takes advantage of the low distortion inherent in laser processing is the welding between the main gear and the synchronizing gear in an automobile transmission. It is possible to maintain the inside radius of the main gear to less than 50 μm (0.002″) and the distortion in the synchronizing gear to less than 130 μm (0.005″). Although the same job could be carried out just as rapidly by an electron-beam welding system, it must be done in a vacuum, and the usual holddown devices (vacuum and magnetic) cannot be used on the workpiece.

Perforating and Drilling

The baby-bottle nipple was one example of using the laser to perforate materials. In another application, a 1-kW CO_2 laser is being used commercially to perforate the plastic upholstery that covers the walls of passenger cabins in jet airplanes. One CO_2 laser manufacturer has developed a special optical tracking system for perforating plastic irrigation tubing as it is extruded at speeds of up to 1.5 m/sec. Lasers have also been used to perforate paper for easy separation into sections. The operation is performed rapidly and cleanly. Beyond the practical advantages of rapid processing of materials, laser perforation can save time and money in maintenance costs. Replacement of the breakable metal perforating pins with a pulsed CO_2 laser has considerably reduced maintenance time for a variety of perforation machines.

Closely related to perforation is drilling. Any distinction between perforation and drilling is in the duration of time or the amount of energy needed to perform the operation. Since most drilling systems operate in a pulsed mode, a number of pulses are need to punch through a material or drill to a specific depth. One of the first applications of the laser was to drill diamond dies used in the extrusion of wire. A ruby laser output was focused onto the center of a diamond tablet and the laser was pulsed repeatedly until a hole of the desired size and shape had formed. Prior to the introduction of this technique, the holes were made by a much slower process using fragile, high-speed drills.

In most drilling applications, a well-controlled hole diameter is desired. Lasers, being noncontacting drilling devices whose machining parameters are determined by the choice of optical components (lenses, masks, mirrors, deflectors, etc.), can be used to drill aerosol nozzles and control orifices with the required precision. One such example requiring two precision-drilling operations is shown in Fig. 9.6. A CO_2 laser is used to drill two metering orifices in a valve at a production rate of 150 parts per minute. The hole sizes can be varied from 0.25 to 1.0 mm in diameter with an accuracy of ± 0.04 mm. One special example in the medical field is the

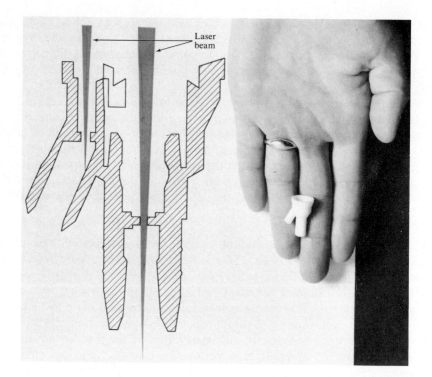

Fig. 9.6 Precision laser drilling of holes in an aerosol valve. A production rate of 150 parts per minute can be maintained using a CO_2 laser to perform the drilling operation. (Courtesy of Coherent Radiation, Inc.)

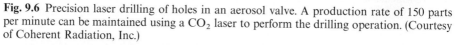

drilling of holes near the edges of hard contact lenses to increase fluid flow across the eye. This increase reduces the irritation caused by the contact lenses for longer periods of time.

Scribing

One important application of the laser in the electronics industry is the scribing of alumina and silicon substrates used in integrated circuits. After a line of easy fracture is laid down on the substrate material by scratching the surface with a diamond stylus or thin metal saw, pressure is applied to break the circuits apart. The replacement of the stylus and saw with lasers has increased the speed and decreased the cost of scribing substantially. Despite scribing rates as high as 35 cm/sec, the precision control of the laser beam permits finer cuts than were previously possible. An example of scribing ceramics with a CO_2 laser is shown in Fig. 9.7. Note that no attempt is made to cut a continuous line in the material. A close-up of a fracture edge, shown in the inset, gives some indication of the size

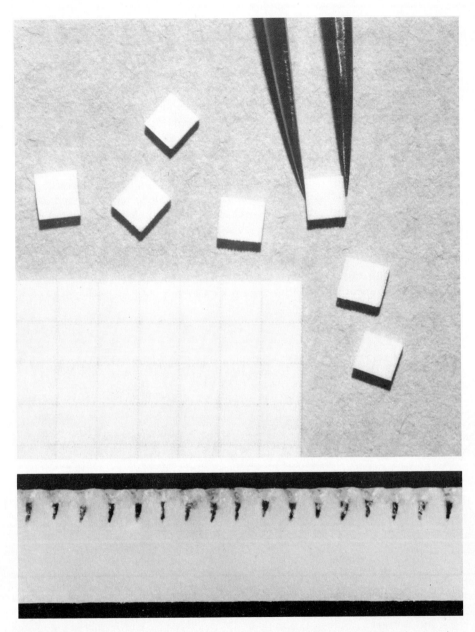

Fig. 9.7 Laser scribing of a ceramic. A CO_2 laser is used to produce a series of craters in a ceramic piece. Provided the holes are close together and deep enough, the material is easily fractured along the scribe line. A close-up of a fracture edge is shown below. The depth of the laser damage, 0.2 mm (0.008″), is a fraction of the total thickness of 0.635 mm (0.025″). (Courtesy of Coherent Radiation, Inc.)

and depth of the scribed craters relative to the ceramic thickness. The precise processing reduces the separation between circuits, thereby allowing more circuits to be packed on a single substrate. Scribing on ceramics is done principally with CO_2 lasers because of the excellent absorption properties of alumina at 10.6 μm. Silicon wafers, on the other hand, are processed with Nd:YAG systems. The Nd:YAG laser is pumped with incandesent tungsten lamps and may be Q-switched with an acoustooptic cell. In a typical system, the pulses are delivered at a rate of up to 10 KHz, with an irradiance of 10^8 watts/cm^2 in an area less than 25 μm in diameter.

Micromachining and Resistor Trimming

Another application for high-power lasers is the machining of extremely small objects. With a tightly focussed laser beam, it is possible to repair printed circuits by machining away short-circuit connections. Photographic masks used to create microcircuits can also be repaired or revised by micromachining. In such applications, the operator must be able to evaluate the work being done and move the workpiece to the appropriate places. In Fig. 9.8, a system is shown that allows the operator to view the system by closed-circuit television. Focusing the image for the television camera with microscope optics also focuses the laser beam onto the

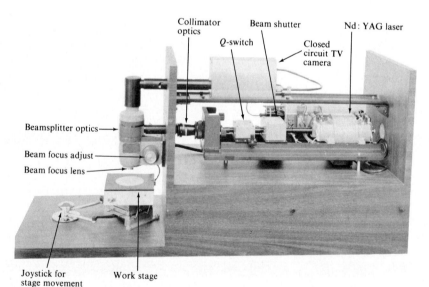

Fig. 9.8 Laser micromachining system. This system incorporates two different features found on many other systems: a closed-circuit television viewing system and an operator-controlled workstage. When in use, the entire apparatus is enclosed in protective shields. (Courtesy of Quantronix Corporation.)

workpiece. The workstage is manipulated beneath the laser beam focus point by a joystick knob. The laser is a Nd:YAG laser that operates 2 watts CW TEM_{00} and can be Q-switched with an acoustooptic shutter. As usual, the entire area is enclosed in a metal cover for safety. All viewing during processing is done via the TV camera.

One of the most widely used laser machining applications is that of trimming precision metal film resistors. A metal film resistor is purposely deposited with a resistance that is lower than the required final value. It is positioned on a stage and two micromanipulators provide electrical contacts to the resistor. A standard current is applied to the resistor and the voltage across the resistor is monitored. The laser beam is focused onto the metal and part of the metal film is vaporized, raising the resistance of the resistor. When the voltage across the resistor rises to a specified value (the standard current times the precise resistance sought), either the operator or a control device turns the laser off or diverts the output from the thin film.

Cutting

One of the most widely known examples of cutting with a laser beam appears in the James Bond movie *Goldfinger*. The secret agent's nemesis cuts his way into the United States gold bullion stronghold at Fort Knox, Kentucky with a laser mounted atop a personnel carrier. Although the film was first released in 1964, the laser in this episode has come to represent the image of a laser in the minds of many. While some industrial lasers today may be equal to the task of cutting through an armored door, these lasers can also cut through a wide variety of materials, rapidly and precisely, and in a somewhat less dramatic fashion.

Starting at the low end of the power scale, lasers are being used to strip insulation from electrical wires and cables. The method is quite effective because the insulation materials used are generally good absorbers of laser light, whereas the metal conductor is a good reflector. The beam from a low-power CO_2 laser (50 watts or less) easily cuts its way down to the conductor without nicking or cutting the conductor itself, maintaining the mechanical and electrical integrity of the cable. For cutting large holes in materials, the focused laser beam is directed in a circular path. Such a path is easily achieved by rotating the focusing lens about an axis parallel to its optic axis. The radius of the hole cut then equals the distance from the lens axis to the rotation axis.

At somewhat higher power levels, glass and quartz are easily cut with a CO_2 laser because of the almost total absorption by these materials at 10.6 μm. Quartz can be worked at room temperature due to its low thermal expansion coefficient. Soda-lime and borosilicate glasses, exhibiting larger thermal expansion at room temperature, must be preheated to around 400°C before working if thermal shock is to be avoided. In a typical operation, quartz tubes of 25 mm diameter or less and wall thickness 3 mm or less can be cut in less than 4 seconds using a 250–500-watt CO_2 laser. The tube is rotated during the cutting operation at 60–120 rpm.

By combining the CO_2 laser with a computer-controlled beam director, intricate patterns can be cut rapidly and precisely in a variety of materials. In a branch of the ready-made clothing industry, for example, the patterns for a large number of models and sizes of dresses and suits are stored on magnetic tape, and the items are cut to order with a laser-cutting system. Synthetic fabrics are particularly suited to this technique, since the cut edges are melted by the beam and any fraying is prevented. There are other cases, such as in the manufacture of cardboard boxes and shoes, for which it is better to use the laser to produce just the master patterns or cutting jigs (the pieces of plywood into which cutting knives are inserted) and then proceed to the normal fabrication process. To improve the cutting of these organic materials, a gas jet coaxial with the laser beam may be added. The gas, usually compressed air or nitrogen, does not take part in any reaction, but prevents burning the top layers of the material and transports the heat generated by the laser cutting deeper into the cut.

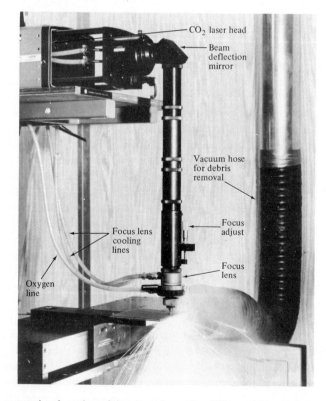

Fig. 9.9 Oxygen-assisted cutting of titanium sheet. The 250-watt beam is deflected downward from the CO_2 laser head mounted above the workstage. (Courtesy of Coherent Radiation, Inc.)

Probably the greatest challenge presented to an industrial CO_2 system is the cutting of metals. For the most part, metals are highly reflective and the cutting action is therefore inefficient. There are methods for improving the efficiency, however. When oxygen is used in the coaxial jet, the gas reacts with the molten metal and increases the cutting rate substantially. Figure 9.9 shows oxygen-assisted cutting of titanium sheet with a 250-watt CO_2 laser. For a low-carbon steel (12 mm thick), oxygen-assisted cutting can be done at a rate of 1.25 cm/sec.

A Cautionary Note

One of the problems of our technological society is determining those jobs or problems that can be solved by using a particular device. This is true for devices as simple as the wrenches used to fix a car or as complex as the electronic computers used to monitor inventory and sales. More to the point here, the idea of putting a laser to work in some task might seem to be the very epitome of modern engineering when, in fact, it could be a very poor solution to the problem. Although a high-power laser can be used to cut thick plywood sheets cleanly, the volume of work, the amount of handling needed, the intricacy and precision of the cut and so on may dictate a commercial power saw costing a few hundred dollars rather than a laser costing thirty thousand dollars. A careful systems analysis is needed to determine if the laser is indeed the best solution to the problem. One should always consider those characteristics peculiar to a laser system:

1. A precisely defined source of thermal energy,

2. Rapid deflection of the beam,

3. Small mechanical forces connected with cutting.

As with any advancing technology, the analysis of a problem may have to be reworked if the cost of the applicable laser system is reduced because of scaled-up production. One way for the engineer and designer to keep abreast of such changes is to read the technical magazines on the subject, including the advertisements.

9.3 LASER FUSION

Millions of years ago, a wealth of energy was deposited in this energy bank of ours called Earth. From the time humankind inhabited the earth, we have been drawing on these deposits, and lately we have been using up our inheritance at an increasing rate. With this increasing demand for more energy, alternative sources must be found. Although short-term solutions such as greater exploration of the earth's resources must continue, long-range planning must include substantially new sources of energy.

One possible solution is to emulate the sun by releasing energy through the fusion of lighter elements into a heavier element. For example, four hydrogen nuclei (protons) can be fused into one helium nucleus, accompanied by the release of positively charged electrons and a large amount of energy. The energy is the result of the conversion of mass to energy according to Einstein's famous equation, $E = mc^2$. For many years, research using high-temperature hydrogen plasma* has been directed toward the goal of producing conditions similar to those found in the sun. Elaborate magnetic "bottles" have been constructed to provide a high plasma density while the gas is being heated. Various improvements have increased the length of time that a plasma can be confined. However, each improvement has come with greater and greater difficulty, and the goal of controlled thermonuclear reaction in confined plasmas is not yet in sight.

High-power lasers may well provide a solution to this problem. Beginning in the early 1970's, laser scientists and engineers and nuclear physicists began working together in an effort to produce the necessary high temperatures and pressures using pulsed laser systems. The basic concept involved is quite simple. A small pellet of frozen heavy water (D_2O) and extra-heavy water (T_2O)[†] is irradiated on all sides with a high-energy pulse from a laser. The rapid heating of the surface of the pellet causes a burning off, or *ablation*, of the outer material of the pellet. This rapid vaporization of the pellet surface is accompanied by a compressional wave that propagates inward to the center of the pellet sphere, as illustrated in Fig. 9.10, squeezing the remaining material to a very dense core. If the compression is sufficiently great, exceedingly high temperatures—in the vicinity of 100 million (10^8) degrees Kelvin—result. Under these conditions, the velocities of the deuterium and tritium atoms are so great that electrostatic repulsion of the positively charged nuclei is overcome and they undergo a fusion reaction:

$$_1H^3 + {}_1H^2 = {}_2He^4 + {}_0n^1 \text{ (14 MeV)}$$

This reaction yields an ordinary helium atom and an energetic neutron, whose kinetic energy is 14 MeV. For many such reactions to take place within the compressed pellet, the high temperature must be maintained for about 1 picosecond, with a material density 10,000 times that of the uncompressed pellet. These conditions require enormous pulse energies from the irradiating laser, and to date sustained reactions have not been successfully initiated. However, pellet compression experiments using lasers have initiated limited nuclear reactions, with neutron yields in the millions.

---

* At very high temperatures, the hydrogen atoms dissociate from their electrons, forming a *plasma*.

[†] Deuterium (D = $_1H^2$) and tritium (T = $_1H^3$) are isotopes of hydrogen, containing one and two neutrons, respectively, in addition to the single proton of the ordinary hydrogen nucleus.

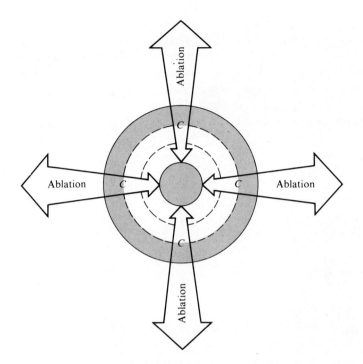

Fig. 9.10 Heating the deuterium-tritium ice pellet causes ablation of the outer surface. The reaction to this outward expansion of material is a compressional wave force (denoted as C) inward.

The lasers presently being used for this research are Nd:YAG and Nd:glass lasers in combination. One stage of research is represented by the Argus laser system, shown in Fig. 9.11, at the Lawrence Livermore Laboratory of the University of California. Starting with a Nd:YAG laser as a master oscillator, the output is amplified by a series of Nd:glass disk amplifiers similar to those shown in Fig. 9.1. The Argus system is designed to deliver from opposite directions two simultaneous pulses of one terawatt (one trillion watts) onto a deuterium-tritium target, compressing the pellet to the high density required for thermonuclear fusion. However, the energy in the beam is not sufficient to generate the high temperatures needed to approach *breakeven*, a condition where the thermonuclear energy generated is equal to the input laser energy. To achieve breakeven, a more elaborate facility, called Shiva, has been designed. In place of the dual chain of Argus, Shiva has 20 amplifier chains, each delivering one terawatt to the pellet from 20 different directions arrayed symmetrically about the implosion

Fig. 9.11 Two-terawatt Argus laser system. A master oscillator (the L-shaped box on the center table) generates a 10-megawatt subnanosecond laser pulse. The pulse moves toward the lower right and is split in two. One pulse moves to the left; the other to the right. At the corners, they are reflected along parallel paths through a series of amplifiers toward the target chamber barely visible behind the wall at the upper left. (Courtesy of Lawrence Livermore Laboratory, University of California.)

point. A model of a 12-arm design is shown in Fig. 9.12. The contorted geometries of some arms are necessary to equalize the distances in all arms, so that the pulses in the amplifier chain, generated from a single pulse in the master oscillator, arrive simultaneously at the target point. The total output of the Shiva system should be on the order of 10^3 joules with peak powers in the range of 10^{13} watts. It is anticipated that an upgraded version of this system could deliver 10^4 joules. This energy, delivered in 100-psec intervals, may be sufficient to prove, at least in principle, that a laser fusion reactor can be a useful power source.

Fig. 9.12 Model of a 12-arm Shiva laser facility. The master oscillator is centrally located in the background of the figure. Each line corresponds to an Nd:glass amplifier chain. In the foreground is the combustion chamber with incoming beams arranged symmetrically about the center. The actual system contains 20 amplifier arms. (Courtesy of Lawrence Livermore Laboratory, University of California.)

Although the neodymium-based laser chain is an attractive initial system for laser fusion studies, it fails on a number of counts. On a practical basis, the solid-state laser cannot be cooled rapidly enough to allow the 100-pulses-per-second repetition rate needed for practical power generation. Furthermore, the numerous flashlamps and disks are expensive and require considerable maintenance. For good coupling to the pellet, the wavelength of the laser should be in the region 300 to 600 nm, not at 1060 nm. Above 600 nm, the absorption of the radiation and subsequent ablation is not as efficient; below 300 nm, two-photon absorption in the disks causes large losses. Unless special fluorine-containing glasses are used for the disks, the losses and the possible damage of components limit the energy flow through the amplifier material to about 1 J/cm^2 per pulse.

Because the excimer laser systems (Section 6.1) do not suffer many of these faults, they are viewed as candidates for the next generation of fusion lasers. The gaseous medium can be cooled quickly by convection and does not require extensive maintenance. Since the excimer laser transitions are between electronic energy levels, some of the excimer laser wavelengths are in the desired 300–600-nm spectral region. Finally, gas, having a lower density than the solid-state medium, will permit energy flow in the neighborhood of 10 J/cm^2. Research has shown that these lasers have an efficiency of a few percent. When developed, it is hoped that the 10 percent efficiency required for a laser fusion pump system can be achieved.

Although the final form of the laser fusion power plant is unknown, present work in the field may give some idea of the shape of things to come. A schematic of such a plan is shown in Fig. 9.13. In much the same way that a soap bubble is blown or a liquid drop is formed, the heavy water ice pellets may be formed as they are needed. An alternative technique uses glass microspheres filled with high-pressure deuterium and tritium gas. The pellet fuel is released into the fusion chamber and, at the precise moment when the pellet is at the center of the chamber, the laser chain is fired. The distances from the master oscillator to the pellet are

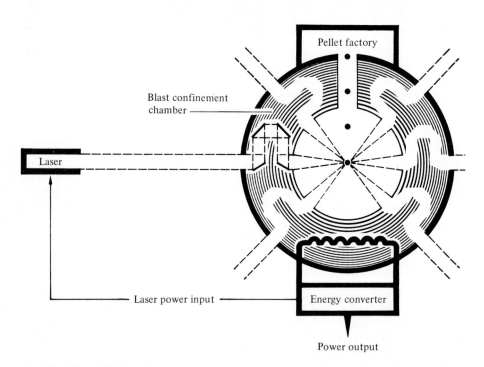

Fig. 9.13 Schematic drawing of a laser fusion power plant. The contorted beam path is designed to admit the laser beam, yet block x-ray and neutron emission from the reaction. (Courtesy of Lawrence Livermore Laboratory, University of California.)

the same for all arms in the chain, so that all the pulses arrive at the center of the chamber simultaneously. (Since light travels at 3×10^8 m/sec, an extra 3 mm of path length corresponds to 10 picoseconds delay.) To admit the laser radiation to the fusion chamber while providing shielding for the x-rays, neutrons, and blast of the fusion reaction, an optical labyrinth of mirrors is used. The energetic neutrons are captured by a lining in the chamber which contains lithium. Lithium is used because it interacts with some of the energetic neutrons to form more tritium. This breeding of additional tritium is necessary for a self-sustaining power system, since there is no natural tritium available. (The original charge of tritium must come from lithium irradiated at a nuclear reactor.) The low atomic mass of lithium $(A = 7)$ permits considerable transfer of kinetic energy from the neutrons $(A = 1)$. Calculations indicate that a hollow sphere of lithium two meters in diameter inside an eight-meter-diameter chamber would serve to moderate the energetic neutrons to the level where radiation damage to the chamber would be considerably reduced. The design of the chamber to contain the blast should provide little difficulty, since the small amount of material involved in the reaction provides a kick no bigger than a good-sized firecracker. The expansion of the ionized particles in the blast against a magnetic field could result in a direct conversion of plasma motion to electrical energy. Additionally, a heated liquid or gaseous neutron moderator could be passed through a heat exchanger to recover energy in a more conventional manner.

The cost of a project such as a laser fusion power plant includes not only the initial construction, but also the recovery and processing of deuterium and tritium plus maintenance of the laser systems. While the fusion process is not 100-percent efficient, the energy return is sufficiently high that research in this field continues at a rapid pace. After all, there is a considerable amount of water in the world! And while the natural abundance of deuterium is low (only 0.0156 percent of all hydrogen isotopes), the potential amount of energy represented by fusion is considerable: Each cubic foot of water represents the energy in 10 tons of coal if all the fusion energy could be converted.

The initial use is expected to be in the simulation of thermonuclear blasts rather than power generation. The large number of neutrons that can be produced at will in a laboratory setting can be used to study the effects of nuclear blasts on critical components, such as the solid-state electronics of offensive and defensive weapons. Such a facility, when fully operable, would reduce the need for underground nuclear tests, which are expensive and hard to instrument. After weapons simulation, other possible applications are neutron conversion processes to produce methane (the major constituent of natural gas) and refurbishing spent nuclear reactor fuel elements. The achievement of a commercially useful laser fusion power plant will probably take much longer to develop than these applications, perhaps to the end of this century.

The Incredible Laser

FOR CREDIBLE
LASERS SEE
INSIDE

Fig. 9.14 Sign found outside many laser research laboratories in the early 1960's. (Courtesy of Professor A. L. Schawlow, Stanford University. The original artist's rendition of the laser cannons appeared in the November 11, 1962 issue of *This Week* magazine.)

9.4 A FINAL POINT

The laser provides an interesting study of the development of a basic device. As we have seen from the preceding material, the many forms and varieties of the laser have been derived from a single device in 1960 and transformed into the key element in a large number of applications today. These applications can be as benign as the helium-neon laser used as an "optical stylus" for playing pre-recorded television presentations on plastic discs similar to phonograph records, or they can be as deadly serious as the laser-directed "smart" bombs used near the end of the Vietnam war. The high power output of the lasers discussed in this chapter could also be used in antimissile weapons (laser cannons) and possibly in antipersonnel weapons. The sign shown in Fig. 9.14 was a jest of the early 1960's; today the footnote could be removed.

As with any new technology or device, the use of lasers to obtain certain goals cannot be taught within the bounds of a textbook. It is up to an informed society and its leaders to decide how lasers will be used. Individually, each one of us must decide whether the application is worthy of our time and effort.

PROBLEMS

9.1 Referring to Table 9.1, design a laser system to perform one of the following operations:

a) Custom-cut wooden jigsaw puzzles.

b) Puncture one million ping-pong balls so that they can be strung on wires. (Believe it or not, one of the authors was approached for consultation on this problem.)

c) Put breathing holes in the edges of hard contact lenses.

d) Removal of a steel entrance door from the United States gold bullion depository at Fort Knox, Kentucky. Please specify your power source and appropriate safety precautions. (Idea courtesy of Ian Fleming.)

e) Rapid labeling of wooden crates by charring the surface. (You will probably need some moving mirrors to deflect the beam rapidly.)

9.2 An atomic mass unit (amu) is equal to $1/12$ of the mass of $_6C^{12}$ and is equal to 1.660×10^{-27} kg. What is the amount of energy that would be released if one amu were converted completely from mass to energy? Express your answer in both joules and MeV's (1 joule $=$ 6.24×10^{18} eV). The table below lists the masses of several atoms and the neutron:

Symbol	Atom/particle	Mass (amu)
$_1H^1$	Proton	1.007825
$_0n^1$	Neutron	1.008665
$_1H^2$	Deuterium	2.014102
$_1H^3$	Tritium	3.016050
$_2He^3$	Helium-3	3.016025
$_2He^4$	Helium-4	4.002603

How much total energy is released in the deuterium-tritium reaction below?

$$_1H^2 + {_1H^3} \rightarrow {_2He^4} + {_0n^1}$$

But the energy of the neutron ($_0n^1$) is given as 14 MeV in the text. How do you explain any discrepancy?

9.3   The total bank energy for a Nd:glass amplifier chain is 3.8 MJ, contained in a glass volume of 80 liters. Assuming identical concentrations of Nd ions for all disks in the chain, what is the concentration, in number/cc, of Nd ions in the disks? (*Hint*: $\lambda = 1.06$ μm for the Nd laser transition.) What assumptions must you make in answering this question?

REFERENCES

Books
9.1   S. S. Charschan (ed.) (1972), *Lasers in Industry*. New York: Van Nostrand Reinhold. One of the best applications-oriented books available. The writing is clear and the examples excellent. An excellent take-off point from this text.

9.2   J. F. Ready (1971), *Effects of High Power Laser Radiation*. New York: Academic Press. A more advanced level of mathematics is used.

9.3   M. Ross (ed.) (1971), *Laser Applications*, Vol. 1. New York: Academic Press. Five applications by five separate authors.

Articles
9.4   A. Ashkin (1972), "The Pressure of Laser Light," *Scientific American* **226,** 62. An article on how small particles can be pushed around by light and some applications that derive therefrom.

9.5   J. L. Emmett, J. Nuckolls, and L. Wood (1974), "Fusion Power by Laser Implosion," *Scientific American* **230,** 24–37.

Index